Making Deep History

Making Deep History

Zeal, Perseverance, and the Time Revolution of 1859

CLIVE GAMBLE

OXFORD

UNIVERSITY PRESS

Great Clarendon Street, Oxford, OX2 6DP,
United Kingdom

Oxford University Press is a department of the University of Oxford.
It furthers the University's objective of excellence in research, scholarship,
and education by publishing worldwide. Oxford is a registered trade mark of
Oxford University Press in the UK and in certain other countries

First Edition published in 2021

Impression: 1

Published in the United States of America by Oxford University Press
198 Madison Avenue, New York, NY 10016, United States of America

British Library Cataloguing in Publication Data
Data available

Library of Congress Control Number: 2020941751

ISBN 978-0-19-887069-2

DOI: 10.1093/oso/9780198870692.001.0001

Printed and bound by
CPI Group (UK) Ltd, Croydon, CR0 4YY

In memory of

Marc Gamble and Drew Morris

Acknowledgements

Roland Pease started this book when he asked me in 2007 to do a talk for the BBC on strange encounters in science. I chose the 1859 time revolution about which I knew little. I was fortunate that three excellent books prepared the ground; Donald Grayson's *The Establishment of Human Antiquity* (1983), George Stocking's *Victorian Anthropology* (1987), and Bowdoin Van Riper's *Men among the Mammoths* (1993). These were followed for the time revolutionaries by Arthur MacGregor's edited volume *Sir John Evans* (2008), Janet Owen's ground-breaking *Darwin's Apprentice* (2013), Mark Patton's *Science, Politics and Business in the Work of Sir John Lubbock* (2016), and John McNabb's insightful *Dissent with Modification* (2012). The task of embedding the time revolution into the nineteenth century was aided by the depth of historical scholarship for this period. Among the books I consulted, three were invaluable; A. N. Wilson's *The Victorians* (2003), David Cannadine's *Victorious Century* (2017), and Jonathan Conlin's *Evolution and the Victorians* (2014). I would never have embarked upon the book without the efforts of the legions of anonymous scanners who, in the last decade, have put so much published material online. Resources such as the *Darwin Correspondence Project* (https://www.darwinproject.ac.uk/) and Mike Pitts's *SALON*, the online newsletter of the Society of Antiquaries (www.sal.org.uk), have transformed writing histories, while *Wikpedia* is the best fact checker the world has ever known. I encourage you all to donate to it.

Four libraries have been essential to my project. At the Royal Society, Keith Moore put me in touch with Prestwich's original manuscripts and referee reports. Heather Rowland, Adrian James, and Bernard Nurse at the Society of Antiquaries found sources and answered questions about illustrators, while Ted Nield and Wendy Cawthorne at the Geological Society granted access to Prestwich's notebooks, and their archivist Caroline Lam uncovered the Fréville Pit photographs from among Prestwich's papers and pointed me to Murchison's archive. The Bibliothhèque Louis Aragon at Amiens is the source for the photographs of the time revolution, and Thomas Dumont showed them to me in the Pinsard archive.

Three museums have been especially important. Jonathan Williams and Jill Cook at the British Museum were generous with information about Henry Christy and the role of their institution in the Victorian period. Alison Roberts at the Ashmolean shared her deep knowledge of Evans and his archive and with Ilaria Perzia allowed me access to the letters and manuscripts. In the Heberden Coin Room, Jerome Mairat located the coin from Amiens that Evans found on the day they demonstrated human antiquity. At the Natural History Museum a special debt goes to Robert Kruszynski. He seized on the project and used his knowledge of Prestwich's collection to bring the discovery flint, where it had been lying forgotten since 1896, back into the light. Phil Crabb of the NHM took its photograph, which we published together in 2009.

Writing history depends on the generosity of others, sharing knowledge, and encouraging completion. I have been especially fortunate, and my thanks go to Lyulph Avebury (information on his great-grandfather, John Lubbock and pictures of Nelly and Alice), K. Paddayya (Robert Bruce-Foote and the Indian Palaeolithic), Pierre Antoine, Émilie Goval, Jean-Pierre Fagnart, and Jean-Luc Locht (the archaeology, geology, and antiquaries of the Somme), Pierre Schreve (deciphering and translating Boucher de Perthes handwriting), David Bridgland (Lubbock's geology), Charles and Sarah Beeson-Jones (Prestwich and London's water supply), John Mather (the scientific importance of Grace Prestwich), Louise Revell (for an honest assessment of the digging abilities of Thomas Wright), David White (Victorian photography), Hugo Lamdin-Whymark (Evans and flint knapping), Chris Evans (archaeology as experiment), Alistair Best (for pointing me to Ruskin), and Tim Pearce for an education that started with Victorian novels.

I am also indebted to those who read the many drafts and put the engine back on the tracks with their comments and encouragement: Janet Owen, John McNabb, Stephanie Moser, Jonathan Gregory, and Alice Wright. And there were three anonymous reviewers for Oxford University Press whose comments helped clarify the text. In addition there have been many productive discussions about the time revolution with Dora Moutsiou, Matt Pope, John Gowlett, David Shankland, Andrew Lawson, Paul Pettitt, Simon Buteux, Peter Middleton, Felix Driver, Dan Smail, Jonathan Conlin, Andrew Shryock, David Bridgland, Chris Stringer, Mark White, Thomas Trautmann, Arthur MacGregor, Rob Hosfield, Chris Gosden, and Elaine Morris.

Crucial work in tracking down sources both in England and France and preparing the text for publication was provided by Jo Hall, Jemma Jones, Cory Stade, and Dan Hunt, while Christian Hoggard organized the illustrations and Kaylea Raczkowski-Wood drew all the original artwork. Kat and Lyn Cutler prepared the index. At Oxford University Press Karen Raith guided the book to publication.

Finally, my debt to two ancestors is acknowledged in the dedication: Marc Gamble, my father and Drew Morris, my father-in-law. Two different, but successful businessmen, possessed of abundant zeal and perseverance, whose unwavering support made my journey into deep history possible.

Contents

List of Figures

List of Tables

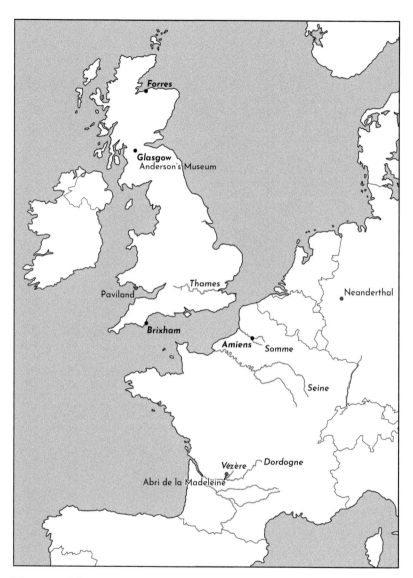

The map of the 1859 time revolution
Source: Kaylea Raczkowski-Wood

What we see depends mainly on what we look for.

John Lubbock, *The Beauties of Nature
and the Wonders of the World We Live In*, 1892: 4

It is impossible not to sympathize with those who, from sheer inability to carry their vision so far back into the dim past, and from unconsciousness of the cogency of other and distinct evidence as to the remoteness of the origin of the human race, are unwilling to believe in so vast an antiquity for man as must of necessity be conceded by those who, however feebly they may make their thoughts known to others, have fully and fairly weighed the facts which modern discoveries have unrolled before their eyes.

John Evans *Ancient Stone Implements* 1872: 622

The more one becomes acquainted with time the more slippery does it appear to be.

Joseph Prestwich in a letter to his
sister Mrs Kate Thurburn, March 1860

Dieu est éternel, mais l'homme est bien vieux.

Jacques Boucher de Perthes,
Antiquités celtiques et antédiluviennes, Volume 2, 1857: 359

1

The Time Revolutionaries of 1859

Prelude

The time revolution of 1859 rang to the sound of splintering stone. It was a swift upheaval, quickly overturning an age-old belief and ushering in a new science. In the space of a few months during a calendar year we became a species immersed in the vastness of time. The old certainties of how and when we originated were damaged beyond repair and, like a cracked bowl, their days of holding water were numbered. Human history now had no time limit, its possibilities transformed.

The year 1859 is, however, feted for a different great upheaval. It was the year when Charles Darwin (1809–82) finally published *On the Origin of Species by Means of Natural Selection*. Its impact was anticipated, immediate, and transformative across all the disciplines investigating the variety of life and its capacity to change. It was accepted and disputed in equal measure. It refused to go away. No scientific book has had so much impact on all aspects of society and culture or received so much detailed analysis. *The Origin*—like Darwin himself—is a global economy of scholarship examining every page and squiggle in his archive, even down to the hairs in his beard.[1]

Darwin casts such a bright light over the year that the time revolution in human history passes largely unnoticed.[2] My purpose is to extract the revolution in human antiquity from Darwin's dazzle and explore the possibilities for human history that it opened up. My exploration follows Thomas Trautmann and his insight twenty years ago that a change in the scale of time necessitated a change in the content of history.[3] He explored the time revolution through issues close to an anthropologist's heart: race, ethnicity, and the acceptance that social evolution took place. I touch on all of these but push to the fore the content provided by human prehistory or, as I prefer, the archaeology of deep human history.

The events of 1859 brought humans and geology together. The idea of deep time was not new.[4] It had been conclusively demonstrated by the international efforts of many geologists culminating, for the British, in Charles Lyell's (1797–1875) *Principles of Geology* published in three volumes between 1830 and 1833—a book which accompanied Darwin on *The Beagle*. Geologists wanted more time to explain how the earth was shaped. That was the 'deep' part of their deep time. But in the absence of scientific measures they were wary about putting an age on anything in either millions or hundreds of thousands of years. Time, instead, was an 'abyss' that represented a 'vast lapse of ages'.

In 1859, humans were not part of this geological time-fest. They were bound instead by a shallow chronology, 6,000 years for the creation of Adam on the Sixth Day—an age arrived at by laborious scholarship over several centuries that added up the ages of elders and prophets in the Bible. This was the Mosaic—the adjective comes from Moses—chronology. In 1859 it dominated not only the discussion of human antiquity but also the historical imaginations of all Victorians. It was an unshakeable age that tied the creation of the earth, its fauna and flora, to the sequence of events recorded in Genesis, where a grand historical design was laid down by a divine architect. The Mosaic account did not recognize natural change, and neither did most Victorians before 1859. Genesis taught that animals and humans were created once. Then they were fixed in time, incapable of change. The idea that they could become something else, another species, due to pressures from the environment was not just unthinkable to most Victorians, but disgusting.

So, to reshape the nineteenth century's historical imagination a wall of words had to be pulled down. It was not enough for the time revolutionaries to point to the profundity of geological time necessary to turn seas into limestone cliffs. What was needed was hard evidence: a humanly made stone tool that would breach the wall of words and unite geology with human history. It was time for an imaginative revolution to sweep aside the barricades of dogma by turning the spotlight of scientific reason onto the question of human origins.

The breakthrough occurred on the afternoon of 27 April 1859, seven months before the publication of *The Origin* on 24 November. It involved two close friends Joseph Prestwich (1812–96) and John Evans (1823–1908), businessmen with a shared interest in geology. Prestwich had been encouraged the year before to turn his geological expertise to the long-standing question, 'How old were humans?' He wrote candidly afterwards that before 1859 'there were [only] twenty men of science in Europe who would have admitted the possibility of the contemporaneity of man and of extinct animals'.[5] Over the course of twelve months, and interrupted by a demanding day job importing wine, his quest took him from a limestone cave above

Brixham in Devon to a gravel quarry outside the northern French city of Amiens. Accompanied in April 1859 by Evans, who was taking a few days away from his papermaking business, and witnessed by local French antiquaries, they found the proof that human history had to be measured by geological timescales. No longer could our past be crammed into the time-depth of recorded history as chronicled in the Bible or the king lists of ancient Egypt. For the first time humans were cut adrift from these certainties. The year 1859 was when history began.

The two scientific revolutions of 1859—deep human time and the evolution of life—fed off each other but were always independent. Darwin did not need the April proof from the Somme to confirm his argument. Humans do not appear in *The Origin*. Neither did Prestwich and Evans require Darwin's mechanism to prove their case that humans coexisted with extinct animals like the woolly mammoth. What linked both of them were the principles of geology. Darwin expressed it succinctly when he wrote in a letter on 5 September 1857, 'In nature geology shows us what changes have taken place, and are taking place. We have almost unlimited time; no one but a practical geologist can fully appreciate this.'[6] It is the idea of *unlimited time* that made change not just possible but probable once its driver, natural selection, had been articulated. Then the principle of descent with modification took hold, or as he put it in *The Origin*'s subtitle, *The Preservation of Favoured Races in the Struggle for Life*. Darwin explained why in the same letter: 'Think of the Glacial period, during the whole of which the same species at least of shells have existed; there must have been during this period millions on millions of generations.' In other words, ample time and population numbers for natural selection to get to work on and shape those 'favoured races'. Furthermore, the glacial period was the most recent in the geological stack of eras and epochs. Looking further down the strata, the number of generations and the length of time available for new life to evolve increased exponentially. Natural selection rather than artificial creation was inescapable.

My purpose in writing this book is to humanize a story of scientific endeavour characterized by abundant zeal and perseverance. Quite simply, there can be no history of how we came to understand our past without an appreciation of the personalities involved and the constraints they laboured under. History as understood and experienced by the individual needs, like our collective history, a dialectic to challenge accepted views; a wall to bowl ideas against.[7] But what deep history also needs is a trampoline to allow the imagination to connect to other pasts: worlds peopled by different humans living by unfamiliar skills. Without the ability to bowl and bounce, deep human history is just a bald timeline shorn of interest. It becomes the servant of ideology with no prospect of contradiction and hence little understanding of how our world has been shaped.

The Victorians were driven by an ideology of progress. There was a purpose to the past and its deep-time component confirmed for them the direction of history. The time revolutionaries and their social networks were responsible not only for putting that train on the tracks but also for giving it a full head of steam. The whistle, first blown in 1859, can still be heard. By contrast, Darwin's vision of evolutionary change is blind, with no directed end point in sight. Few grasped this essential feature of his theory as it applied to humans. And it has taken more than a century to wrestle deep history away from the tramlines of inexorable historical advance.

The wall I am bowling against in this book is built from the objects that allow us access to our deep past. The time revolution of 1859 relied upon a proxy, a stone tool, to stand as evidence for the human ancestors who left no words and only a few of their skeletons behind. And here is the paradox at the heart of the 1859 time revolution. These stone proxies are very old, while the words used by the time revolutionaries in their letters and papers to build their case are extremely recent. Here is a dialectic juxtaposing not only timescales but also the types of evidence, objects and words, available to the historian. The deep past may seem impersonal because stone tools do not talk. I aim to show in this book that they do, however, have a voice, amplified more than a 150 years ago by a small band of time revolutionaries who conclusively proved they were historical objects. Ever since they have turned the key in the door of deep human history.

There was nothing inevitable about 1859's time revolution. Throughout, the vagaries of personal histories emerge as more important than adherence to the storyline that scientific truth, whatever that might be, would eventually be revealed. The history of discovery is a messy business. It is commonly depicted, however, as inspired, if not heroic. It was messy because these constellations of people and their scientific passions were enmeshed in the social, economic, and political worlds of the mid-nineteenth century. The time revolutionaries had their own desires, motives, and ambitions. They spun connective webs from the threads of family ties, the conduct of business, as well as their leisure pursuit of gentlemanly science. As a result, the threads in the time revolution's history are tangled and sometimes difficult to unravel. The events surrounding 1859 are highly compressed from our perspective, but a chronology is still discernible; it runs from a day

(Chapter 2) through a month (Chapter 3) into a year (Chapter 4) and ends with two bites at the decade (Chapters 5 and 6). I have used the historian's trick of breaking with the calendar to create a long month and decade. The legacy of these time-shaping events is set out in Chapter 7, where the story is brought up to date. The events that led to a revolution were largely forgotten by archaeologists during the intervening years, and these are resuscitated in an Afterword (Chapter 8).

If I was to dress up this 'messy' history in its academic Sunday best, it would be as an exploration of *entanglement*:[8] our propensity to create meaning by wrapping or bundling experiences together. These bundles can include disparate things like an ancient stone tool and a Victorian paper magnate. Wrapping incorporates ideas, people, and objects from different places and varied ages. In this way, 'time', as Bruno LaTour remarks, 'is always folded'.[9] The ancient rubs against the modern, and a simple view of time as linear, while useful for the narrative arc of a time revolution, obscures the historian's task, which is to unwrap the bundle of time.

As an archaeologist I excavate such time bundles. I am interested in objects not just as historical markers but also as material metaphors—concepts to think with, without words. This unthinking exercise is possible because metaphors are grounded in our daily experiences.[10] Material metaphors, artificial objects, have been around much longer than human language. Their antiquity and the changes their forms underwent suggest a way to link the present with the deep past and so learn something new about our remote ancestors.

Today our metaphorical imaginations are dominated by objects which contain our bodies and lives: the clothes we wear, the cars we drive, and the houses we live in.[11] So much so that we do not give them a second thought. Such containers also dominated the conceptual world of the 1859 time revolutionaries with their carriages, gentlemen's clubs, and the notion of compartmentalized, separate spheres. But this was not the conceptual world of deep human history that the time revolutionaries were about to uncover. Things that contained, or wrapped, if you prefer, other things are rare in the Old Stone Age. The imaginations of our earliest ancestors were dominated instead by objects with different metaphorical possibilities. These were handheld instruments, one of which, the stone hand-axe, would be the focus of the time revolution of 1859. This is the simple object, very old and very abundant in the Old World, at the heart of my book. I wrote it to find out what happened when time is folded and two conceptual universes collide. There are no surprises that the time revolutionaries fashioned the remote past to their way of thinking, and we have followed their lead ever since. By writing this history I hope, however, to clear a way to understanding our deep history that avoids recreating the past with dominant metaphors from the present. The 1859 time revolution was a necessary product of its age. That much is certain. But in shaping the scientific decades that followed, it took on an ageless, post-revolutionary quality as succeeding generations fed the interest in deep human history with new content. We continue to do this in order to negotiate our relationship not just with remote ancestors but with time itself. This is where the archaeologist and historian combine and, to let them flourish, for the rest of this book their Sunday best will remain under lock and key in the academic wardrobe.

Water and Unlimited Time

Darwin's statement that only practical geologists can appreciate the existence of unlimited time is apt when we consider Prestwich and Evans.[12] They met on a train in 1854. They were bound for the Richmond Assizes to appear as expert geological witnesses for opposite sides in the Croydon water question. This was a long-running dispute about the ownership of London's underground water supply.

The case concerned the River Wandle, a south-bank tributary of the Thames, where a mill owner claimed that the well he had dug did not infringe the rights of the Local Board of Health to the underground water. The case hinged on 'a thing so obscure and uncertain in its existence as underground water'.[13] In short, mill owners wanted the right to augment their water supplies irrespective of the fact that it might inconvenience others. The initial judgment went through two appeals, reaching the House of Lords in February 1859. Their Lordships finally dismissed the mill owner's case on 27 July.[14] Prestwich saw a practical benefit to his pioneering geological work on the water sources of the London Basin.[15] Evans was disappointed. He needed the water from the River Gade to make paper at Dickinson's mills at Abbots Langley in Hertfordshire. When accused of being a polluter of the water supply, he made a typically robust answer, 'I am not ashamed of my occupation,' hurling geological arguments back at his critics.[16]

Water was a big political issue in the 1850s. The Thames was a public disgrace culminating in the Big Stink in the summer of 1858, when Members of Parliament sweltered behind closed windows because the stench was so foul. The river was a killer, blind to rank. Evans survived typhoid in 1852,[17] the disease that killed Prince Albert at Windsor in 1861.[18] Letters to *The Times* complained about the 'pestilential mud banks' that appeared at low tide around Old Westminster Bridge.[19] Others, who today would be called 'stink deniers', noted a decrease in the smell during 1859 and a resulting improvement in the appetites of Thames watermen, evidence they used to exonerate the river as the cause of a substantial increase in cases of diarrhoea, one of the symptoms of cholera, across the city.[20]

Eventually MPs could hold their noses no longer. They found the money to start work on the great infrastructure project that built London's sewers and pumping stations and created embankments where the new underground railway was hidden. The source of the funding held an irony for Prestwich because it came from duties on wine and coal.[21]

When the two men first met, they passed the journey from London in silence. On arrival in Richmond they found the assizes had been cancelled that day, and on the train home they forged their geological bond. Throughout their friendship over the next forty years those initial interests of looking deep underground and mapping watercourses, past and present, would constantly recur. In April 1859 the interests came together as they searched, in a paraphrase of the Croydon water question, for 'a thing so obscure and uncertain': a stone implement made by someone who lived alongside extinct ice age beasts. This was the proof they sought to answer the question, 'How old was Mankind?' And to answer it they applied their zeal and perseverance.

Principals and Partners

In 1859 the bespectacled and luxuriously bewhiskered John Evans was 36 (Figure 1.1). He was of medium height and stocky, with long, delicate fingers that fluttered excitedly when he handled antiquities, especially coins, which were another of his passions.[22] Short-sighted he might have been but he was above all things *steadfast*.[23] This quality won him the accolade from Thomas Huxley (1825–95) as someone who 'knows the wickedness of the world and does not practise it'.[24] But with such perseverance came a tendency for long-windedness and, as he grew older, pomposity.[25]

When he was 16, he had been sent to work at Abbots Langley in the highly successful paper business John Dickinson & Co. Ltd.[26] Through very

Figure 1.1 (a) Joseph Prestwich (1812–96) date unknown. His wife Grace drew attention to his fine forehead. In his obituary Evans wrote 'we have lost not only one of the great pillars of geological science, but a geologist whose mind was as fully stored with accumulated knowledge as that of any of his contemporaries'.
Source: G. A. Prestwich 1899.

Figure 1.1 (b) John Evans (1823–1908) *c.*1855 just before he took over the management of John Dickinson's, the paper manufacturer. When Evans died in 1908, Lord Avebury (John Lubbock) recalled that 'in private life he was a delightful companion, a genial host, and a staunch friend'.
Source: Evans 1943.

hard work and ability he climbed the company ladder, taking over its running from his uncle, John Dickinson, in 1856.

John and Harriet, his Dickinson cousin, married in 1849, when he was 26 and she was 23 (Figure 1.2). Their marriage was strongly opposed by her parents, and even after her father allowed it, but before he handed on the business to John, she saw her living standards drop. The newly-weds lived in Dickinson's brick-built Red House, 'small, inconvenient and ill-placed',[27] positioned deliberately in sight of the lodge gates of the grand, but empty Nash House and overlooked by the Dickinsons' forbidding new mansion on Abbots Hill.[28] This deeply unsympathetic bully paid John so little that Harriet had to abandon Madame Amable's French fashions and make do with bonnets made from local straw.[29] Her photograph as mistress of Nash

Figure 1.2 (a) Harriet Anne Dickinson (1826–58), John Evans's cousin and first wife, *c.*1857 the year before she died at the age of 31. She was survived by five children: Arthur, Lewis, Philip, Alice, and Harriet.

Source: Evans 1943.

Figure 1.2 (b) Frances (Fanny) Phelps (1826–90), John Evans's cousin and second wife in a painting by Jane Mary Hayward in 1859, the year of their marriage. When travelling, Evans wrote many letters to her that began 'Dearest Child'. They had no children. Fanny was devoted to Harriet's children. Her gift of music was wasted on John.

Source: Evans 1943.

Figure 1.2 (c) Maria Millington Lathbury (1856–1944) Evans's third wife. On their marriage in 1892 Prestwich wrote to him, 'There may be some disparity of years but you may well take twelve or thirteen off yours, seeing how vigorous you are in mind and body. Besides, the disparity disappears before a parity in tastes and thought.' Their daughter Joan was born a year later.

Source: Evans 1964.

House in 1857 is a study in quiet determination; she is wrapped in a swirl of expensive fabrics that proclaims those days of straw hats are now over.

Harriet gave birth to a daughter on 19 December 1857 at Nash House into which they had moved two years previously. All was well with mother and child, but the nurse they hired carried an infection which Harriet contracted and to which she succumbed ten days later.[30]

Five children were left motherless, three boys and two girls. The oldest, Arthur, was 6 on New Year's Day 1858. Because of their mother's illness the children were sent to their grandparents at Abbot's Hill. When they returned, John wrote in his dead wife's diary that they did not seem to feel her loss, which was crassly insensitive but unexceptional from a Victorian father. Arthur Evans came across his parents' diary seventy years later and scrawled an angry *NO* in the margin.[31]

John regarded his eldest son as an odd child. In February 1859, after a dinner party where he met William Thackeray the author of *Vanity Fair*, he wrote to his cousin Fanny Phelps—and here the wheel spins round again, because he had just got permission to marry her from their uncle, John Dickinson—that he does not understand Arthur. The small boy, who will uncover the Minoan civilization of Crete, had broken the legs of a china doll and buried the remains in the garden. And above the grave Arthur placed the inscription 'KING EDWARD SIXTH and the butterfly and there [sic] cloths and things'.[32] Steadfast John liked facts and was mystified at this behaviour in an Evans. 'The Psyche element is very singular,' he wrote, 'and the placing of the clothes in readiness for his re-existence looks like forethought.' But was it so very singular? A sensitive child had lost his mother and was left with a father whose attention was so in demand elsewhere that he was ill-equipped for the emotional demands of the life domestic. By 1859 it was time for John to marry again. Fanny returned from Madeira on 8 July, and two weeks later they were married at St George's, Bloomsbury. Their fortnight's honeymoon was 'sternly archaeological',[33] a subject about which Fanny had a dutiful but limited curiosity. John's zeal for antiquities, however, overrode any other concerns:

> They went to Colchester and examined Dr Duncan's[34] collections and the gravel pits at Lexten; they went to Norwich and looked at the Cathedral, the Castle and the Museum; they relaxed a little among the architectural beauties of Ely and Peterborough, though at Peterborough there were gravel pits too. They continued by York to Scarborough,

where Jonathan Rashleigh[35] turned up and they saw Mr Bean's[36] collection of fossils; they went by Lincoln (and more collections) and Nottingham (with a little business thrown in) to Bosworth, and so home to Nash Mills'.

Evans's close friend Joseph Prestwich, scion of the family firm of Joseph Prestwich & Son, importers of wine and spirits at 20 Mark Lane in the City of London,[37] was older by eleven years. He began working in the family firm at 18, and while the business made him rich, it was, as another geologist observed, full of 'uncongenial mercantile cares'.[38] In 1859 he was unmarried, but his wife Grace (they tied the knot in 1870) later described his principal feature as an unusually fine forehead. Joseph was driven and determined about the truth. 'Nothing', Grace Prestwich wrote in 1899, three years after his death, 'stirred his indignation so much as when he met what was false, or a sham, or underhand, and then he spoke out his mind.'[39] And he spoke with a booming voice from a thin, tall body.[40] Prestwich's zeal for scientific order and accuracy, one geologist noted, was balanced by 'the perennial charm of his personality' and the 'gentle child-like simplicity of his heart, his unaffected modesty, and his genuine goodness'.[41] In short, as his memorialist summed him up with a misspelt line from Chaucer, 'He was a verray perfight gentil knight.'[42]

Such delicacy meant he was often torn between supporting some theory to spare someone's feelings, while knowing that geologically it was wrong.[43] There was, also, a touch of imperiousness in his make-up.[44] When he was confronted by pretension or affectation, his response was to label that person as possessing 'a great deal of manner'.[45] He was serious-minded, founding the short-lived Zetetical Society when he was 21.[46] The name speaks to the search after truth, its members dedicated to the cultivation of scientific and literary knowledge. Self-improvement was its goal. His photograph captures the highbrow and nervous intensity that led to bouts of ill health, as well as a set of spreading whiskers that no modern cheeks could be without in the 1850s.

Joseph was tightly wound. In the photograph his watch chain is especially prominent; his right hand poised to whip out his watch from his fob pocket. He lived by the timetables in *Bradshaw's Railway Guides*, essential to organizing his business and geological journeys. 'Travel, travel, travel', were Sir Charles Lyell's three pieces of advice to all geologists, and they drove Prestwich.[47] Evans recalled that Joseph's commercial trips around Britain aided rather than restricted his geological studies.[48] He was a constant visitor to France and Belgium, and his fluent French allowed him to build up a network of geological contacts as well as visiting gravel pits and geological sections exposed in railway cuttings.

Prestwich was the eldest son in a large family of seven children who survived infancy.[49] The youngest of his five sisters Civil Mary (1822–66) took over his household and became his scientific amanuensis. Following their father's death in 1856, the two siblings moved into the family's elegant Regency town house, designed by John Nash, at 10 Kent Terrace, Regent's Park. At Kent Terrace, Civil rescued her brother from an ascetic bachelor's life and organized his scientific work. She transcribed rough manuscripts in his looping hurried hand to send to scientific journals. She kept a register of his geological finds and drew up a book of bibliographical references which, according to Grace, was a model of method and order.[50] This database recorded Prestwich's wide geological interests. The book had four ruled columns for England, France, Europe, and other parts of the world. In each column were listed the authors and their scientific papers on Eocene, Miocene, and Tertiary geology.

Joseph formed an important friendship with the Scottish botanist, palae-ontologist, and geologist Dr Hugh Falconer (1808–65); a vivid character who wrote lively, humorous letters, unlike Prestwich and Evans, and who took up academic causes (Figure 1.3). A generous scientist, Falconer was always keen to cooperate and share his extensive scientific knowledge: 'Underlying all the glee and laughter-provoking sallies there was the deeply affectionate and genial nature which drew Joseph Prestwich as with a magnet.'[51] In a satirical sketch about evolution, 'A Sad Case', published in 1863 in *Popular Opinion*, the two friends were described for its middlebrow readership as 'Falconer, the old bone man' backed up by 'Prestwich, the gravel sifter'.[52]

Falconer had worked for many years as an economic botanist in Pakistan where he oversaw the commercial introduction of tea from China into Bengal.[53] At the same time, he collected fossil animals in the Siwalik Hills, and found the first bones of a fossil ape. He hit the headlines with the discovery of a gigantic fossil tortoise (Figure 1.3) that in his opinion under-pinned the Hindu mythology of a world tortoise.

Returning to England in 1855 for health reasons, he became interested in the question of human antiquity and evolution. Darwin regarded him as one of his oldest friends, and he made the sage of Down House laugh: 'What a splendid joke about Falconer,' exclaimed Darwin in a letter sent on 9 February 1859,[54] although sadly the joke has not survived.[55] Falconer could tease Darwin without offence being taken. In a letter a month before

Figure 1.3 An 1868 cartoon of the giant tortoise *Colossochelys atlas,* found by Falconer and Captain Cautley in 1835 in the Siwalik Hills, and shown carrying the Elephant and the World. Falconer believed the giant fossil could have existed alongside the earliest humans in India. These speculations began his interest in the question of the antiquity of mankind. The sketch, by geologist Edward Forbes (1815–54), shows the elephant wearing glasses, as Uncle Hugh did, and holding his book on the fauna of the Siwalik Hills.
Source: Falconer 1868: 377, Fig.13.

The Origin appeared in bookshops, he described it in a postscript as 'the wicked book which you have been so long a-hatching'.[56] For his part Darwin could write to him as 'my good old friend and enemy'.[57]

Falconer's niece, Grace Milne (1832–99), who later became Prestwich's wife, was an important witness and participant in the time revolution (Figure 1.4). Born near the town of Forres in Morayshire, she married George McCall in 1854, and her son James was born a year later.

Figure 1.4 (a) Grace Anne (Milne McCall) Prestwich (1832–99). This image by W. E. Miller, based on a pencil sketch, is dated 1876 and was published in her memoir compiled by her sister Louisa Milne in 1901. There are no known photographs of her.
Source: Milne 1901.

Figure 1.4 (b) Dr Hugh Falconer (1808–65) graduated from Edinburgh as a medical doctor in 1829 and became assistant surgeon in Bengal for the East India Company. He returned because of ill health in 1855.
Source: G. A. Prestwich 1899.

Heartbreakingly, both her husband and infant son died within two months of each other in 1856.[58] She suffered depression, writing that she felt 'so crushed' and lying 'half waking, half musing on the listless waste of my aimless life, thinking that I was a useless weed in the garden of well-growing plants and trees', only supported by a will superior to her own that revealed the glory of her blessings.[59] She was strong-willed and devout, but her recovery also needed the timely intervention of Uncle Hugh, who in September 1858 invited her to join him on his winter journey to the sun of the French Riviera, Italy, and Sicily. 'I do not like the idea of travelling alone,' he wrote, 'and it has occurred to my obtuse head that if you would accompany me it might be a pleasant arrangement for us both.' But there were conditions, 'Only two gowns: when the old one wears out buy a new one. No band-boxes! Write like a good girl, and say Yes.'[60]

In 1861 she moved into her Uncle Charles's elegant house in 21 Park Crescent, London, which on his return to England Hugh Falconer, his brother, had also made his home. Here she presided over the household, acted as hostess to scientific parties, and was a valued note-taker. Some of her scientific illustrations survive, and she published her own works on geology and Scottish history.[61] Five years after Falconer's death in 1865 she married Prestwich, twenty years her senior. Park Crescent is a twenty-minute walk from Kent Terrace.

The bond between Falconer and Prestwich strengthened when the former became involved in the excavations at Windmill Hill Cavern[62] above the small fishing town of Brixham in Devon. Quarrying the limestone hillside in January 1858 revealed deep fissures containing bones of ice age animals, including hippopotamus and elephant.[63] The importance of the cavern was seen immediately by William Pengelly (1812–94), an experienced geologist living in nearby Torquay. Falconer visited and subsequently urged the Geological Society of London to establish a committee for its investigation to resolve palaeontological issues. This was approved, and Prestwich became its Treasurer.[64] Funds for excavation were obtained from the Royal Society and Pengelly put in charge of supervising the workmen on a daily basis aided by a local subcommittee based in Torquay. What began as an investigation into the geological history of ice age animals changed on 29 July 1858 with Pengelly's discovery of stone implements. These were duly reported,[65] but not published, by Pengelly and Falconer at the 1858 autumn meeting of the British Association for the Advancement of Science in Leeds.

Early in 1859, when Falconer and his niece were in Sicily, Prestwich wrote to him about progress at Brixham.[66] He reported that excavations had been ongoing since the previous year, only stopping for a week at Christmas. William Pengelly was distracted by a sick daughter, and another member of the Brixham Committee, Robert Godwin-Austen (1808–84) was now con-vinced that the flint instruments occurred in association with the bones of extinct animals. 'After my last visit I cannot deny it,' Joseph wrote, 'but still I am not satisfied without seeking every other possible explanation besides that of contemporaneous existence.' Never closed-minded, but inherently careful, Prestwich confided to Falconer, 'None of the evidence which has come before me during the last ten years has appeared to me conclusive.'[67] Two months later he would have conclusive proof from Amiens.

The last and youngest of the practical geologists and time revolutionaries was the polymath John Lubbock (1834–1913). In 1859 he sported a bushy beard and the first signs of a receding hairline (Figure 1.5). He knew Prestwich, Evans, and Falconer through the Geological Society.[68] Lubbock's star had risen quickly. He had been elected a fellow of the

Figure 1.5 (a) John Lubbock (1834–1913) in 1867. His obituarists described him as a 'busy man', amiable and urbane, who invariably came down on the side of the 'soft answer'. 'Useful' was the word that summed him up.

Source: © National Portrait Gallery, London.

Figure 1.5 (b) Nelly (Ellen Frances) Hordern (1836–79) in 1856. His first wife, she took an active role in his scientific pursuits and published papers on the archaeology of Scandinavia. John and Nelly had six children.

Source: Courtesy of Lubbock Family Archive.

Figure 1.5 (c) Alice Fox-Pitt (1862–1947) second daughter of General Pitt-Rivers, taken shortly after their marriage in 1884. They had met the year before. She was twenty-eight years his junior. They had five children.

Source: Courtesy of Lubbock Family Archive.

Royal Society in 1858 at the age of 24.[69] This was in recognition of his work on insects. Those who signed his ballot from personal experience included Prestwich, Thomas Huxley, and Charles Babbage (1791–1871)[70] as well as Charles Darwin, his mentor and neighbour to his childhood home in Kent. Beetles were not his only interest. Antiquities, particularly those from Scandinavia, ethnology, and geology all excited his curiosity, and cricket was his sport. His interests led to a union of natural with human history, and the juxtaposition of undirected Darwinian evolution with well-intentioned Victorian progress.

At this time Lubbock was working in the family bank of that name. There is a description by Dickens in his bestselling *A Tale of Two Cities*, serialized in 1859, of the institutional ageing process for someone like Lubbock in a City bank; 'When they took a young man into Tellson's London house, they hid him somewhere till he was old. They kept him in a dark place, like a cheese, until he had the full Tellson flavour and blue-mould upon him.'[71] The young John, however, would never acquire the full Tellson flavour. He would be a cheese-maker, not a cheese.

Plain John became Sir John in 1865, when he assumed the hereditary baronetcy on the death of his father.[72] A glittering career in banking and politics was predicted, although Darwin desperately wanted him to stick with science.[73] He entered Parliament as the Liberal MP for Maidstone in 1870 and was ennobled Baron Avebury thirty years later.

In 1854 he married Nelly, Ellen Frances Hordern (1836–79), and in 1859 their first two children, Amy Harriet and John, were in the nursery at High Elms in Kent, the large, crowded family home. Nelly accompanied him on many archaeological visits in the UK and Europe. Her vivacity and striking beauty attracted the notice of the Danish antiquary Jens Worsaae.[74] Like Civil Prestwich and Grace McCall, later Grace Prestwich, Nelly Lubbock not only acted as hostess but also read the proofs of his papers, prepared diagrams for his lectures, and contributed her own paper on the ancient shell mounds of Denmark to the volume *Vacation Journals* in 1862.[75] John's scientific friends confided discreetly in their journals that Nelly was her husband's superior in both sensitivity and force of character.[76]

Lubbock's world added another sphere of influence to those of John and Joseph. Initially this layer was supplied by Darwin and his wide scientific circle. Later it would extend to Lubbock's political life. During 1859 Darwin was completing *The Origin*, and he drew on Lubbock's knowledge about insects. They had known each other for more than a decade. As a young boy, John would take his butterfly net and his latest specimens and walk from

High Elms to Down House, where Darwin had been living since 1842. Darwin's children showed less interest in their father's bugs and fossils, so the regular visits from a studious and serious 14-year-old naturalist were welcome. Charles even encouraged John's father to buy the boy his first microscope.[77] Neither was he John's lone cheerleader. Sir Joseph Hooker (1817–1911), Darwin's closest friend and scientific confidant, eminent botanist and Director of Kew Gardens, wrote of Lubbock in 1863: 'I think him the most faultless character I know, who is at the same time one of the best and most active and clever. Thank God there are some men worth living for.'[78] Praise indeed from a master of the caustic phrase, who went on in the same letter, 'but really when one peeps beyond the immediate circle of one's friends one gets disgusted and disquieted altogether'.

Scientific Circles

Friends and close scientific colleagues were critical to the success of 1859's time revolution (Figure 1.6). Their affection and trust powered a common enthusiasm to demonstrate that human history had a geological rather than biblical time-depth. At the heart of a web of networks was the circle of friendship between Evans, Prestwich, and Lubbock. They shared many interests. All three were fellows of the Geological Society and the Ethnological Society, which became the Anthropological Institute in 1871, and, after Evans was elected in 1864, the Royal Society. Lubbock and Evans were further tied by the Society of Antiquaries.

Fellowship and friendship went together. In particular, there was an aura among geologists that arose, as historian Roy Porter described it, from a 'proclaimed zest, capacity for physical endurance, unfailing good humour, honesty, bluffness and larger-than-life qualities'. Geologists, he went on, were living proof 'that geology was an activity of spirit: it was soul food'.[79] Lubbock, Evans, and Prestwich were fully paid-up members of this geological banquet, and its unwritten code sealed their bonds of friendship.

In the next circle were three important gatekeepers to other networks vital for the time revolution; Darwin, Falconer, and the French antiquary Boucher de Perthes (1788–1868). Darwin and Lubbock were close, the social and scientific reach of John's mentor immense, as revealed by his correspondence.[80] Darwin's opinion was sought on many scientific matters, human antiquity among them, and was disseminated through his regular

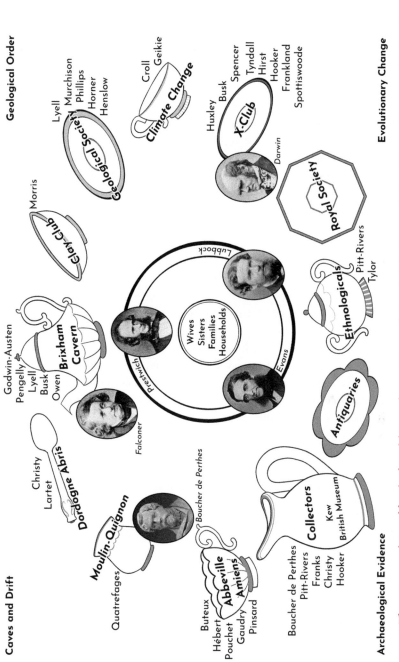

Figure 1.6 The spinning plates of family, friendship, and acquaintance that circled the time revolution. The result was a unification of knowledge where natural and human history were drawn together at different timescales.

Source: Original artwork Kaylea Raczkowski-Wood.

letters to Joseph Hooker, Thomas Huxley, and a host of others across Britain, Europe, and North America. Hugh Falconer, ever the mover and shaker, was the gateway to French scientists, initially geologists and palaeontologists, but eventually the isolated antiquary in the northern town of Abbeville, Jacques Boucher de Perthes. They were introduced in 1856. Boucher de Perthes had the information they sought and a devoted local network of supporters among the learned societies of Abbeville and Amiens. What he lacked was any backing outside the Somme for his claim that he had found evidence for the time revolution in that valley.

Clustered around the time revolutionaries and these three gatekeepers was a constellation of scientific endeavours, clubs, and events that form, like spinning plates, the story of the time revolution. Some, like the investigation of a limestone cave at Brixham drove the quest for evidence. Others, like the 'X-Club'[81] or the circles of Collectors and Ethnologicals, sought to order and interpret the new content for a new time-depth.

Separate but Integral Spheres

Within the time revolutionaries' inner circle were their wives, sisters, families, and households. These are too easily overlooked. The reason is simple; the networks of the time revolution were inescapably dominated by males. The Victorian creed of separate spheres which divided male from female and work from home was respected like a set of rails.[82] But male scientists, not to mention men of business, received crucial support from wives, sisters, and nieces who in comparison to their loud, whiskery brothers, uncles, and husbands formed an influential, if historically less vocal sisterhood.[83]

This binary system of separate spheres extended to the mind. The historian Henry Buckle (1821–62) proposed in 1858 that male minds were inductive, driven by facts, and female minds deductive, open to general ideas.[84] This explained for him why men made scientific discoveries. Women, however, played a vital role in tempering rigid male induction with imagination and enthusiasm. In this way, he argued, women rendered a great, though unconscious service to science.[85] Without them scientific progress would be slow and ponderous. The type fossils for Buckle's curiously gendered minds could have been Hugh Falconer and his niece Grace McCall. Falconer stratified his mind into two layers; the older was inductive, the younger, deductive. The reason he invited Grace to join him on his trip south in 1858 was that he found himself too prone to following his deductive mind as it leapt to conclusions. As a result his inductive mind was 'growing old and stupid', and he required someone to watch over his declining years.[86]

Occasionally we see what managing the separate sphere of the household involved while the men, unburdened by domestic crises, were off hunting antiquities. April 1861 and Fanny, at Nash House, was having a particularly 'dreary time with a dying cook'.[87] But then, 'J[ohn] came in unexpectedly by the late train in great spirits having found two beautiful flint implements at Reculvers.'[88] He was followed soon afterwards by Prestwich and Sir Charles Lyell.[89] Fanny did not tell her diary who cooked their Sunday dinner.

Three women assiduously husbanded the posthumous reputations of their men. In the year she died Grace completed the life and letters of her husband Joseph Prestwich, a self-confessed labour of love.[90] Unsurprisingly, she presented a portrait of her husband that was highly flattering, even saintly. The historian Dame Joan Evans (1893–1977), John's youngest daughter by his third marriage, published *Time and Chance* in 1943; a detailed, loving memoir of her father, whom she knew only as an old man. The biography mostly chronicles her eldest half-brother Arthur, destroyer of dolls and discoverer of Knossos. Through these memorials both women hold the reins of memory from which this history of the time revolution is written. Alice, Lubbock's second wife, never wrote an account of her much older husband. But she did selectively edit his voluminous papers, destroying those she regarded as sensitive. In particular, she redacted his diary where he expressed his feelings for Nelly, his first wife.[91]

The sisterhood may have been muted in 1859, but it was never, like the next stratum of essential support, invisible. These were the servants who drove the carriages, banked the fires, cooked, washed, ironed, and emptied the chamber pots. We encounter them for a fleeting moment every ten years, when the UK census reveals who was under which roof.

On census night, 7 April 1861, Joseph and his sister Civil had two live-in servants at 10 Kent Terrace; Mary Read and Elizabeth Painter, both 31 years old. At Nash House, Abbotts Langley, John and his second wife Fanny listed nine, including a governess, cook, footman, and maids. The most crowded house was High Elms at Downe in Kent, where Lubbock's parents lived. John was there with Nelly and their young family. Altogether there were fifteen Lubbocks present, looked after by eleven live-in servants. Falconer and McCall missed the census because they were in Italy.[92]

There are no records that Evans and Prestwich travelled with valets and personal servants who carried their bags, freed their hands, and allowed them to concentrate on the business in hand. That was how John Fowles in *The French Lieutenant's Woman*, set in 1867, imagined the pairing of his fictional gentleman geologist, Charles Smithson, and his manservant Sam Farrow,[93] someone who according to Wilkie Collins, writing in 1859, 'was in that state of highly respectful sulkiness which is peculiar to English servants'.[94] If Collins's opinion was indeed the general rule, then John and Joseph would have gladly travelled alone. Just occasionally the servants break through the baize door of invisibility. Evans's butler William Hemmings wrote a diary of great dullness that incorrectly suggested his master was a Freemason.[95]

Ladies' maids did accompany their mistresses, as Grace Prestwich revealed in the account of her travels in 1858–9 with Uncle Hugh. She engaged an Italian lady's maid, Carolina Belloni, in Paris for their excursion to Italy.[96] They got on well. Nelly Lubbock's maid, Annie, was with them on the Birmingham train that derailed in 1865. Nelly and John, up front in first class, almost died. Annie at the back in third was unscathed.[97]

Zeitgeist and Time in 1859

These stratified, entangled networks, so important for making the time revolution of 1859 happen, were enfolded in the mood of the year at the end of a prosperous, optimistic decade known as the Age of Equipoise, the Victorian heyday.[98] Two of 1859's rich crop of new books capture the year's zeitgeist. These are Samuel Smiles's *Self-Help: With Illustrations of Character, Conduct and Perseverance* and George Eliot's *Adam Bede*. Both were on John Lubbock's list of his 100 favourite books.[99]

By one of those calendar coincidences that pepper 1859, *Self-Help* was published on the same day as *The Origin*, 24 November. Smiles's book, the first of thousands of future titles about self-improvement and lifestyle choices, dealt with momentous changes to the individual on a very rapid timescale. Not for Smiles those millions of generations of shells that Darwin had described in his vision of unlimited time in 1857.

When it came to accounting for change, it was the character of the individual that mattered to Smiles. *Self-Help* sketches the lives and achievements of many great men, and all his examples are male, from all walks of life. These were either self-made men or, if they started with advantages, they built on them. Evans fits the former Smilean type, Prestwich and Lubbock the latter.

Smiles lists six qualities of character that make a successful businessman. These are Attention, Application, Accuracy, Method, Punctuality, and Despatch. The three time revolutionaries demonstrated these attributes in their papermaking, banking, and wine businesses. They had the perseverance and zeal to succeed. In today's language they were both strivers and achievers. Of the three, Evans had the greatest struggle to establish himself in his unpleasant uncle's thriving paper mill. But once established and wealthy, he adopted the motto, worthy of Smiles, 'I desire to deserve'.[100] Joseph would have agreed with Smiles's firm opinion that 'the person who is careless about time will be careless about business' and not to be trusted, therefore, with important matters such as the antiquity of humans.[101] In his later writings Lubbock set out to outdo Smiles with uplifting aphorisms for the self-made man: 'Never waste anything, but above all, never waste time,' he declared in 1894, and went on to outdo Cicero by stating that 'Self-confidence is no doubt useful, but it would be no more correct to say that what was wanted was firstly perseverance, secondly perseverance and thirdly perseverance.'[102] Lubbock's earlier blockbuster, *The Pleasures of Life*, published in two parts in 1887 and 1889, is a paean to the century's virtues: 'people can generally make time for what they choose to do; it is not really the time but the will that is wanting'.[103]

Time was a nineteenth-century preoccupation, in particular, its power to shape nations, business, politics, history, architecture, arts, and crafts.[104] This power lies at the heart of Darwin's work with his insistence on slow, gradual change leading to new life forms. By contrast Smiles shows with his chosen scientists, politicians, and engineers that the journey from log cabin to White House can be fitted into a single breathless lifespan. The capacity for change either lies outside an individual's control (Darwin) or, with the correct application, is firmly in their grasp (Smiles). Darwin's sense of time, which he learned from the geologists, has the power to shape life among populations. Smiles took a different view. Populations did not interest him. His manual of self-help puts the individual above the herd. These tall poppies strive to get ahead and in his world are justly rewarded for using their talents. He wrote, 'man perfects himself by work more than by reading,—that it is life rather than literature, action rather than study, and character rather than biography, which tend perpetually to renovate mankind'.[105] This was the creed of the practical geologist. Channelling Smiles, the Scottish science writer and geologist David Page (1814–79) declared in 1869, 'A day well spent in the field is worth a dozen of reading at the fireside.'[106]

Smiles's huge bestseller presented an aspiring zeitgeist; success was the reward of a moral character and hard work. And the converse: if you were successful, it was obviously because you were morally upstanding. So, while Evans and Prestwich were about to follow Darwin and embed ancient humans in unlimited amounts of time, their understanding of change and history owed more to the timescale of individual endeavour promoted as the ideal by Smiles. They adhered to the mantra set out by the public intellectual Herbert Spencer (1820–1903) in 1858 that everything—life, history, culture, society—proceeds from the simple to the complex.[107] And that was enough for them to consolidate their proof for great human antiquity.

The power of time to shape the contemporary mood is one of George Eliot's themes in her novel *Adam Bede*, which went on sale on 1 February. It was Eliot's first literary success. In two years the novel went through nine editions, its sales vindicating Marian Evans's pseudonym, which survived much speculation until July 1859, when she was unmasked.[108]

Eliot started her novel very precisely on 'the eighteenth of June, in the year of our Lord 1799'. A new century beckoned for a new type of man. By 1859 these men and their wives, her readers, had their feet firmly under its table. Her main character, Adam Bede, is a village carpenter known for his skills, industry, and integrity. Eliot took as her inspiration Samuel Smiles's earlier book on *The Life of George Stephenson* (1781–1848), one of the greatest of engineers and known as 'The Father of the Railways'. It was first published in 1837, and in it Smiles holds Stephenson up as a beacon of liberal values: humble origins, hard work, morally sound judgements and a believer in free trade, small governments, and self-reliance. Stephenson emphatically ticked all six of Smiles's criteria for a successful businessman, and Eliot poured some of these attributes into Adam.

In doing so Eliot collapsed time and folded, like a dust-jacket, the zeitgeist of her readers around her characters. The pace of change in those compressed sixty years introduced a new social type, the man of New Leisure. In one passage Eliot steps away from her story to directly address her readers, who, while familiar with the technological marvels driven by steam, still hankered for a whiff of golden-age nostalgia.

Leisure is gone, gone where the spinning-wheels are gone, and the pack-horses, and the slow wagons, and the pedlars, who brought bargains to the door on sunny afternoons. Ingenious philosophers tell you, perhaps, that the great work of the steam engine is to create leisure for mankind. Do not believe them: it only creates a vacuum for eager thought to rush in.[109]

And it is the men of New Leisure who rush in. Lubbock, Prestwich, and Evans had eager thoughts to fill the vacuum such as the scientific pursuit of butterflies, the geology of London, and the discovery of Primeval Man. The outcome, as Eliot saw it, is that:

> Even idleness is eager now—eager for amusement; prone to excursion-trains, art museums, periodical literature, and exciting novels; prone even to scientific theorizing and cursory peeps through microscopes.

The industry and drive of Lubbock, Evans, and Prestwich stands in marked contrast to Eliot's Old Leisure who was:

> a contemplative, rather stout gentleman, of excellent digestion; of quiet perceptions, undiseased by hypothesis; happy in his inability to know the causes of things, preferring the things themselves. Life was not a task to him, but a sinecure. He fingered the guineas in his pocket, and ate his dinners, and slept the sleep of the irresponsible.

Here was someone who belonged to a previous zeitgeist and had not benefited from reading Samuel Smiles. Eliot knows what she is doing, when, having set up Old Leisure for a fall, she then lets him off the hook: 'Do not be severe upon him, and judge him by our modern standard.' That was the past, and *we* are now better than that.

How much better is shown by the wonders of nineteenth-century ingenuity. Distance as a proxy for time was constantly being shrunk by the power of steam and the speed of the telegraph. On 25 April that year, and to the roar of cannon, the French diplomat and entrepreneur Ferdinand de Lesseps (1805–94) struck the dusty Egyptian soil with a pickaxe. Work on the Suez Canal had begun, and ten years later it opened for shipping.[110] An enthusiastic report of this ground-breaking event in *Le Monde illustré* declared that now there were no limits to human power: 'Distance is no more... the desert is no more. No more do the ages hold secrets for mankind.'[111] The Suez Canal shortened and made safer the voyages from Europe to the Indian Ocean and then beyond to new colonies in Australasia and the Pacific. The reduction in distance was more than 4,000 miles, the saving in sailing time measured in months. The blow from de Lesseps's pickaxe came at the end of a decade which saw the ruthless onslaught on time by the agents of modernity: steam, electrical impulses, and the capture of light.[112]

In Britain, the clock which is the symbolic heartbeat of the nation began ticking on 26 April, the day Evans set off to meet Prestwich in France. The chimes of its bell, Big Ben, took a little longer. They first rang out on 11 July, making Londoners look up to check the time. Such icons of public time at the heart of government were rolled out across the nation as time was standardized to make the railways run.[113] On 2 May Brunel's bridge across the Tamar was opened by Prince Albert. Cornwall was now joined to England, but not without resistance from the magistrates of Penwith. They objected to the imposition of uniform time set in Greenwich and symbolized by Big Ben and the Euston Station clock.[114] Their solution, widely ridiculed, was that Cornish clocks should have two minute hands to show local *and* railway time.[115] Cornwall might have been behind the times, but this ingenious timepiece briefly satisfied the different pace of Old and New Leisure.

The railways created a space for George Eliot's New Leisure that was filled by reading and reflection. Novels such as Dickens's *A Tale of two Cities* were serialized and devoured as the countryside flashed past. Like the coal for the engine, they fed week by week the temporal vacuum created by travel; time, speed, and the space to read all coming together in James Tissot's painting of a prosperous gentleman, warmly wrapped, clutching his watch, strap-hanging to cope with the speed of travel and with a book to pass the time.[116]

A standard time was a small battleground for local identity. The larger war was between technology and Nature. Marvels of the age such as the steam clipper *Royal Charter* shrank the distance between Liverpool and Melbourne, reducing it to an unbelievable sixty days at sea. However, on the night of 25 October the storm of the century, named after this ship, wrecked her on the Anglesey coast less than a hundred miles from port. Only forty-one survived from more than four hundred on board, and with them went a fortune in bullion from the goldfields of Victoria. The body of Captain Wither, a passenger, was washed ashore, and according to *The Illustrated London News* his watch 'had stopped at half-past seven'.[117] Efforts redoubled in the Meteorological Department within the Board of Trade to provide shipping forecasts, the challenge overseen by Darwin's former captain on the *Beagle*, Admiral Robert Fitz-Roy (1805–65). Every day the nation's weather stations telegraphed him with their data on barometric pressure and wind direction. These were analysed under FitzRoy's eye to provide rudimentary forecasts. In the year that human antiquity was pushed back into the 'vast lapse of ages'[118] there were others seeking to drive forward into the future to change the course of history.

Nature played havoc with distance-shrinking, time-saving communications. A series of gigantic solar flares between 27 August and 7 September disrupted the telegraph. This was the Carrington event, named after the amateur astronomer Richard Carrington who identified unusual sunspot activity through a brass telescope in his garden in north London. It is still the largest recorded magnetic storm, captured over twelve days as jagged spikes on paper magnetograms by Greenwich Observatory.[119] The storms knocked out telegraph lines in Europe and North America. Some operators even received burns as their equipment arced fire from the circuits.

In 1859, time was being attacked on the battlefield of Solferino, south of Lake Garda in northern Italy. The French-Sardinian victory over the Austrians on 24 June concluded a phase of the Italian Risorgimento.[120] More than 250,000 troops were involved, and casualties, dead and wounded, amounted to 39,000.[121] The carnage was described in shocking detail by the Swiss businessman Henri Dunant as 'a European catastrophe':[122] 'If the new and frightful weapons of destruction . . . seem destined to abridge the duration of future wars, it appears likely', he wrote, 'that future battles will only become more and more murderous.'[123] Conflicts were shortened not only by advances in artillery, such as the rifled field gun that the French deployed at Solferino,[124] but also by their use of railways to move thousands of troops with astonishing speed to the plains of Lombardy.[125]

New technologies shaped creative imaginations; in particular photography's ability to capture light and freeze time. It was a boast of the Lubbock family that John was the first person in England to be photographed.[126] Louis Daguerre (1787–1851), the French photographic pioneer, captured him when he was a little boy of 5.[127] The picture was destroyed by fire in 1967.

Charles Dodgson, better known as Lewis Carroll, had a passion for photographing celebrities and children. On a visit to the Isle of Wight in April 1859 he visited Alfred and Mrs Tennyson at Farringford. Proofs of *The King's Idylls*, as Dodgson called them, were strewn around the Poet Laureate's study.[128] In Dodgson's album was a photograph of the 6-year-old Alice Liddell as a half-clad beggar child, taken in the summer of 1858 in the deanery garden at Christchurch, Oxford. The wild-haired poet with rotting teeth[129] pronounced the Alice photographs as the most beautiful he had ever known.[130] And in an exchange between two of the most fertile, but different creative imaginations of the century, Tennyson told Dodgson that he often dreamed long passages of poetry which he entirely forgot in the morning. 'You, I suppose', he said turning to Dodgson, 'dream photographs?'[131] The reply from the creator of the Jabberwock is not recorded.

Time and history were prominent in the popular literature of 1859, where the pace of modernity was countered by a romantic tradition with a strong retrospective element. *Idylls of the King*, set in an Arthurian past, was a huge success. The mood of time's relentless advance was captured by Edward FitzGerald in a stanza from his bestselling translation of the *Rubaiyat of Omar Khyam*:

> The moving finger writes; and having writ,
> Moves on; nor all thy Piety nor Wit
> Shall lure it back to cancel half a Line,
> Nor all thy Tears wash out a Word of it.[132]

This journey, unstoppable as an arrow, was described by Dickens in his historical novel of chaos and terror, *A Tale of Two Cities*. Throughout, he used carriages to embody the inexorable momentum of history.[133] As the novel moves through the Terror that followed the Revolution of 1789, the carriages of the rich are transformed *by* time into the tumbrils that carried them to the guillotine. In the process Dickens drew a distinction between history and magic. History leads to irreversible change, while magic can reverse this process of metamorphosis. Butterflies can become caterpillars once again, and Alice can shrink and grow depending on which bottle she drinks from.[134] Hence Dickens's impassioned cry: 'Six tumbrils roll along the streets. Change these back again to what they were, thou powerful enchanter, Time, and they shall be seen to be the carriages of absolute monarchs, the equipages of feudal nobles.' But, of course, this cannot happen. 'No; the great magician who majestically works out the appointed order of the Creator, never reverses his transformations. . . . Changeless and hopeless, the tumbrils roll along.'[135] Who controls time controls the direction of historical change. This was what his readers, now calibrating their lives to the chimes of Big Ben, wanted to hear.

The Task Facing the Time Revolutionaries

What obstacles faced our two earnest businessmen, Prestwich and Evans, in achieving a revolution in time and adding an unimaginable depth to human history?

In front of them stretched a big ditch.[136] Before the time revolution of 1859 this conceptual gulf separated the deep time of geology from the shallow time-depth of Genesis. The age of geological time could be measured in millions of years. In contrast the Mosaic chronology was calculated from the Book of Genesis as taking 6,000 years.[137] By 1859 a widely held compromise had emerged where humans were found only on the Mosaic side of the ditch and all extinct animals and the remote geological eras on the other. This compromise recognized a special time for humans and separated their divine origin from the rest of the animal kingdom (Table 1.1).

Another significant time marker existed within the Mosaic chronology. This was between the antediluvian and post-diluvian, before and after the great deluge. Noah's Flood provided a powerful story of watery global catastrophe. It accounted for the survival and extinction of creatures great and small. It also wiped the slate clean and reset the clock so that history could start again. Only Noah and his kin squeezed through the bottleneck of

Table 1.1 Chronological divisions used in 1859

Mosaic history from Genesis	Boucher de Perthes 1847–57 Flood catastrophe	Prestwich and Evans 1859 Geology of the Drift	Lyell 1839–63 Geology of a single ice age	Lubbock 1865 Christy 1865 Stone Age periods
Creation account				
6,000 years timescale	Thousands of centuries	10,000 years	100,000 years	100,000 years
Children of Noah	Post-diluvian	Recent	Holocene	Neolithic (Surface)
The Flood	Geological flood	Action of rivers and ice	Action of rivers and ice	Action of rivers and ice
Adam and Eve	Antediluvian	Post-Pliocene	Pleistocene	Palaeolithic (Cave & Drift)
Pre-Adamites	Geological eras and epochs, Pliocene/Preglacial			

The geological terms used in 1859 can be confusing because they were unstandardized and used interchangeably by different authors. These are the main ones used in this book. Mosaic is the adjective from the Old Testament prophet, Moses. Holocene and Pleistocene are the geological terms used by earth scientists today and were coined by Lyell in 1839 in an appendix to the French edition of his *Principles of Geology* (Berggren 1998: 124). On the basis of the shells the deposits contained he subdivided the Drift, in ascending order, into the Pliocene (Newer), Pleistocene (Newest), and Holocene (Entirely New) in which we live in today. Estimates of the length of time involved in human antiquity were highly speculative (Van Riper 1993).

the Flood, the sole founders of an unbroken history to the present day. The origins of the world's races were traced by many Victorian scholars to each one of Noah's sons, tillers of the earth and herders of sheep. Even before the Flood the biblical story began in a world with a fully formed domestic economy that drew its staples from farmed crops, flocks, and herds. Consequently, there was no place on this side of the big ditch for those, such as the indigenous Australians and Inuit, who in the nineteenth century lived by hunting and gathering wild foods.[138] They were explained away as peoples who had degenerated from the civilized, agricultural ideal, an argument supported by where they lived—a long way from Paris and London at the uttermost ends of the earth.

But while Genesis was used to organize knowledge about foreign peoples,[139] its Mosaic history was not immune to the overwhelming evidence of geological deep time. By 1859 it had made at least one important accommodation. Geological time was now filled by the pre-Adamites, ancestors with a special nature who lived before Adam and Eve.[140] But there was still no place for ancestors of so-called 'savage' peoples who lived by hunting. And it was out of the question that humans were linked through a chain of evolution to other creatures in the animal kingdom. The big ditch kept the separate spheres, human and animal, farmer and hunter well apart.[141]

What Evans and Prestwich did in 1859 was fill in the big ditch and unite the recent with the remote. To do this they backfilled the ditch with archaeological evidence from limestone caves and the geological Drift. The latter had caught Prestwich's attention. The onomatopoeic Drift mantled old rocks such as the Chalk found widely in southern England and northern France.[142] This covering consisted of gravels and sands that pointed to an origin from floods, rivers, or ice. These sands and gravels also contained the bones of cold-adapted species such as the woolly mammoth and woolly rhino—their woolliness known from carcasses preserved in the Siberian permafrost and brought back to St Petersburg for scientific study.

By 1859 there were claims that humans lived during the ice age at the same time as these extinct beasts. But they were either ignored or laughed at. This was the case for Jacques Boucher de Perthes, who published the first volume of his massive treatise *Antiquités celtiques et antédiluviennes* in 1847. He was the reason that Evans and Prestwich were heading to Abbeville and Amiens in April 1859.

Prestwich was one of the first geologists to take the Drift seriously as the place to find evidence for great human antiquity. The goal was to find humans alongside those extinct animals. That association would be enough to fill the ditch and take humans into the uncharted waters of geological time. Evans and Prestwich started their search knowing it was unlikely they would discover human remains. They were hoping instead to find a proxy that proved an ancient, antediluvian human presence. This proof boiled down to accepting stone implements as human handiwork. Convincing the sceptics that a roughly made stone tool was indeed human rather than natural required attention to detail and a compelling set of arguments. Prestwich's experiences with Brixham Cave pointed to the difficulties of aligning the various strands of geological and archaeological evidence. Conviction alone was not proof, a situation that Boucher de Perthes had discovered.

Filling the ditch was their immediate task. But what possibilities for a deep human history did this act unleash? What searches for scientific content did they set in motion to fill these vast historical timescales?

In the first place Prestwich and Evans added a deep basement beneath the three ages of prehistoric archaeology devised by Northern Antiquaries.[143] John Lubbock would call this the Palaeolithic, Old Stone Age.[144] They filled it with new types of stone artefacts that, once recognized, and properly described, were soon found everywhere. The widespread evidence for such an early Stone Age opened up the possibility of a universal history for humankind that applied to all inhabitable countries. The Palaeolithic became a historical stage from which all human societies had evolved. Its stone tools, so numerous at Abbeville and Amiens,[145] became a proxy for the human bones which proved so hard to find. Indeed, Evans and Prestwich's only involvement with human fossils came in 1863 when they investigated the jaw found by Boucher de Perthes at Moulin Quignon outside Abbeville.[146]

A second spin-off to their time revolution was the interest now generated in the geology of the Drift and the framework for a Pleistocene ice age. Prestwich and Lyell believed in a single period of cold and ice, but soon evidence accumulated which pointed to a more complex picture of repeated advances and retreats. Climate changed from warm to cold and back again, and James Croll (1821–90) proposed an astronomical mechanism to explain these regular cycles with alternating conditions.[147] Prestwich was sceptical, Darwin enthusiastic, as such changes could aid the process of speciation by natural selection as climate isolated populations.

Other lines of inquiry interested Evans the antiquary to some extent but not Prestwich the geologist. They were, however, explored by John Lubbock, who used the time revolution to rethink the place in history of the stone-using hunters and gatherers—the savages of Victorian anthropology.[148]

With the longer timescales they became, in Lubbock's stadial prehistory, examples of progressive social evolution and not degeneration from a once better state. Their inclusion in an upward historical trend gave the optimistic idea of human progress so dear to Victorian values and Lubbock in particular a very deep history indeed.

Less attractive to both Prestwich and Evans, but seized on by Lubbock after 1859 was the link the *unlimited time* of the time revolution made between humans and Darwin's evolutionary process that applied to all other animals. This was the most contentious result of their ditch-filling. The separate spheres containing animal and human were broken. Huxley would thrillingly combine the elements when he argued for continuity in structure between apes and humans,[149] while Darwin's *The Descent of Man* in 1871 would finesse the argument for human evolution.[150] The races were no longer set as original, traced back to Noah's sons like fingers on a hand. Instead there was a unity to mankind based on a single common origin from which all visible variety had evolved under a process of natural selection. Darwin's views were resisted, and Noah's sons still have their adherents.

Prestwich and Evans made their great discovery in the same year as Darwin's *The Origin*. But unlike Lubbock neither was an ardent evolutionist. Evans did discuss the similarity between the development of coin types and the law of natural selection.[151] But he did not apply the same reasoning to flint tools. Prestwich's stance on evolution was closer to Lyell's, although they disagreed about aspects of the principle of uniformity whereby the present is the key to the geological past.[152] All agreed with the great geological law of superposition, adumbrated by Charles Babbage, 'that the order of succession of strata indicates the order of their antiquity, the lowest being always the oldest'.[153] The problem was the Drift geology containing human antiquity, because how it was formed was rarely so straightforward.

Neither did the time revolutionaries set out to defy religious teaching. Evans and Prestwich were practical men who, while passionate in their scientific interests, prided themselves on being free from theories about the past. This freedom allowed them to concentrate on what they did best—standing as expert witnesses whose judgement could be trusted because they stuck to the facts. They knew from the outset that a chronology of 6,000 years was simply bad geology, just as they knew that the sun did not go around the earth. They were the counterpoint to their friend John Lubbock, who pursued the logical consequences of the theory of natural selection, applied it to human history, and by so doing removed the last obstacle to understanding why Adam had ancestors.

Notes

1. Hughes 2017.
2. Trautmann 1992: 381. His paper 'The Revolution in Ethnological Time' emphasizes the speed with which Prestwich and Evans's proof was accepted and how others built this into the Darwinian agenda of change through social evolutionism.
3. Trautmann 1992: 384.
4. Rudwick 2005.
5. Prestwich 1864b: 213.
6. *Darwin Correspondence* Darwin to the American botanist Asa Gray (1810–88), 15 September 1857, in which he sets out his theory of natural selection. This was reprinted in a paper presented to the Linnean Society of London on 1 July 1858 (Darwin and Wallace 1858: 51). This patchwork paper presented the idea of natural selection as put forward by Alfred Russel Wallace, whose letter in 1858 setting out his views had panicked Darwin. The joint paper established Darwin's prior claim, gave Wallace due recognition, and provided the spur for Darwin to get on and complete *The Origin*. Neither author was present when the paper was read out. Wallace was in Papua New Guinea and Darwin was unable to make the short journey up to town from Kent. The paper and the communication were brokered and presented by Charles Lyell and Joseph Hooker (Ashton 2017: p. 3061).
7. I first encountered the 'wall to bowl against' in Peter Wilson (1988: 2).
8. Thomas 1991; Hodder 2012; Pauketat 2013.
9. LaTour (2005: 201; Serres and LaTour 1995: 60) provides an invigorating analysis of how and why we should always think of time as folded rather than linear or cyclical.
10. Lakoff and Johnson 1980.
11. Gamble 2007; Tilley 1999; Knappett 2005; Malafouris 2013.
12. Huxley (1863: 113) also asserted that 'geologists are right'.
13. *The Times*, 17 February 1859, p. 10.
14. *The Times*, 28 July 1859, p. 11.
15. In 1851 Prestwich published a seminal account of the water-bearing character of the geology of the London Basin and in 1857 his book, *The Ground Beneath Us*, dealt in detail with the geology of Clapham, where he was born, and London more generally.
16. The critic was the chemist Sir Edward Frankland (1825–99) in a letter to *Nature*, 16 March 1876, p. 392. Evans's reply was published on 30 March 1876, p. 425. We will meet Frankland in the X-Club in Chapter 6.
17. Evans 1943: 84.
18. Typhoid fever was the cause given on his death certificate, although other illnesses have been suggested: http://theconversation.com/what-really-killed-prince-albert-85939, accessed 24 July 2020. The Prince of Wales recovered from typhoid in 1872 (Cannadine 2017).

19. W. H. Brakspear, Letter to *The Times*, 6 October 1859, p. 9.
20. William Ord, St Thomas's Hospital, Letter to *The Times*, 11 August 1859, p. 9. The miasmic (bad-smelling) theory of how disease spread was still the majority view in 1859; among its supporters, was Florence Nightingale. Foul water was only gradually recognized as the vector for spreading disease.
21. Halliday 1999: Kindle loc. 3282. Prestwich was a member of both the Royal Coal Commission (1866) and the Royal Water Commission (1864), as well as being a busy wine merchant based at 69 Mark Lane in the City of London.
22. Evans 1943: 57.
23. Evans 1964: 21.
24. Evans 1943: 157.
25. Gamble and Moutsiou 2011: 57.
26. Evans 1943: 57–60.
27. Evans 1943: 83.
28. Abbott's Hill was designed by his uncle and built from the grey stone railway sleepers on which the London North Western Main line had once run (Evans 1943: 58).
29. Evans 1943: 88.
30. Evans 1943: 93.
31. Evans 1943: 94.
32. Evans 1943: 99. Edward was the sickly son of Henry VIII who reigned briefly and died at the age of 15.
33. Evans 1943: 105.
34. Dr P. M. Duncan was one of the founders of the Colchester Museum and a local collector of antiquities.
35. Rashleigh was an old friend of Evans and a fellow numismatist.
36. William Bean's fossils can be seen in the Whitby Museum, https://whitbymuseum.org.uk/whats-here/collections/fossils/, accessed 24 July 2020.
37. The Post Office London Directory, 1843: 334.
38. Woodward 1899: 380.
39. G. A. Prestwich 1899: 21.
40. Evans 1949: 121.
41. Archibald Geikie in G. A. Prestwich 1899: 420.
42. Woodward 1899: 381. Line 72 in Chaucer's *Prologue to the Canterbury Tales* reads 'He was a verray parfit gentil knight.'
43. Evans 1949: 121.
44. Harrison 1928: 199.
45. Evans 1949: 121.
46. G. A. Prestwich 1899: 38.
47. Porter 1978: 820–1.
48. *Proceedings of the Royal Society of London*, 1896, 60: xii.
49. His parents had ten children (G. A. Prestwich 1899: 9).

50. G. A. Prestwich 1899: 102.
51. G. A. Prestwich 1899: 195.
52. G. A. Prestwich 1899: 181. The sketch by Dr Pycroft of Exeter was published in *Public Opinion*, 2 May 1863, pp. 497–8, and poked fun at the scientific controversy between Huxley and Owen. There are lots of jokes about monkeys. The dispute between Falconer and Lyell (see Chapter 5) is hinted at in the sketch, which is why Prestwich is named as Falconer's backer.
53. Murchison 1868: 30.
54. *Darwin Correspondence*, Darwin to Lubbock, 9 February 1859.
55. Tantalizingly, for *1859*, where jokes from the principals are non-existent, that page of Lubbock's letter is missing. John Evans's many letters, as his daughter Joan discovered, are lucid statements of fact and common sense. Sadly for her, they lacked any leisurely self-reflection that a biographer can dig into. Instead, 'all his flights of fancy, his criticism and appreciations, his jests and his tenderness, were cast on the running waters of speech' (Evans 1943: 110).
56. *Darwin Correspondence*, Falconer to Darwin, 25 October 1859.
57. *Darwin Correspondence*, Darwin to Falconer, 17 December 1859.
58. Mather and Campbell 2007: 252.
59. Milne 1901: 9.
60. Milne 1901: 12.
61. Milne 1901: 52; Mather and Campbell 2007: 254–5.
62. The site has long since vanished and is normally referred to as Brixham Cave (see J. Prestwich 1873b).
63. For a full account, see J. Prestwich 1873b; Gruber 1965; Grayson 1983: 179–85; Van Riper 1993: 74–116; Wilson 1996.
64. Besides Falconer (Chairman and Secretary), Prestwich (Treasurer), and Pengelly (Director), other notables on the scientific committee included Charles Lyell, George Busk, Richard Owen, and Robert Godwin-Austen.
65. Cook 1997: 118.
66. G. A. Prestwich 1899: 120, Letter, 4 February 1859, Prestwich to Falconer.
67. G. A. Prestwich 1899: 120, Letter, 4 February 1859, Prestwich to Falconer.
68. Lubbock was elected to the Geological Society in 1855, and Evans, proposed by Prestwich, in 1857. Prestwich had been a fellow since 1833 and Falconer from 1842.
69. Owen 2013: 19.
70. A scientific polymath, Babbage is credited with the idea of the first computer, a 'difference engine' that performed mathematical calculations.
71. Dickens 1859 [2000]: 57.
72. Thompson 2009: 8.
73. *Darwin Correspondence*, Darwin to Lubbock, 11 June 1865, on reading *Pre-Historic Times* published that year: 'I do sincerely wish you all success in your election & in politics; but after reading this last chapter [on savage life] you must let me say Oh dear Oh dear Oh dear.' Lubbock was not elected to Parliament until 1870.

74. Owen 2000a: 64. The Irish physicist and friend of the Lubbocks John Tyndall fell in love with her in a manner described by Barton (2018: Kindle p. 4528) as 'structured by Victorian mores'.
75. Owen 2000a: 64.
76. Barton 1998: 440. The friends were John Tyndall (Journal, 22 July 1862, 12 November 1864) and Thomas Hirst (Journal, 22 November 1863 and October 1879 after Nelly had died); both were members of the X-Club; see Chapter 6. After her death these friends felt free to comment on Lubbock's superficiality (Barton 2018: Kindle p. 9095).
77. Owen 2013: 5.
78. *Darwin Correspondence*, Hooker to Darwin, 24 May 1863. Flandreau (2016) presents an alternative picture of Lubbock as a perpetrator of 'white collar crimes'. These involved his using anthropological information to further the aims of the Company of Foreign Bondholders in which he was variously chairman and banker. In Lubbock's defence, he was a successful Victorian capitalist with all the downside that brings in postcolonial times. But the case that he orchestrated a sustained conspiracy between capital and anthropology is weak. Patton (2016: Kindle p. 226) provides a measured account.
79. These were the 'brethren of the hammer, knights errant, a spiritual confraternity in search of a stratigraphical grail' (Porter 1978: 819).
80. https://www.darwinproject.ac.uk/, accessed 24 July 2020.
81. See Chapter 6.
82. Cannadine 2017.
83. Branca 1975.
84. Buckle 1858b: 399. A plethora of self-help books have subsequently traced the differences between men and women to innate differences in biology, intellect, and emotional and behavioural predispositions. They usually enjoy a brief popularity before the artifice in the opposition is unmasked. Buckle was writing at a time when women were allowed neither to vote nor to go to university. This imbalance appears to have eluded him, and most of his contemporaries, as he described the world he knew.
85. Buckle 1858b: 400. In similar fashion, Huxley distinguished between the female mind that accumulated knowledge and the ability of the male mind to generate new knowledge (Malane 2005: 52). The separate mental spheres dominated Victorian thinking.
86. Milne 1901: 12, Letter from Falconer to McCall, 25 September 1858.
87. Evans 1943: 114. Evans 1863: 64 provides a description and illustrations of the two unpolished stone axes he was so pleased with.
88. Fanny is referring to the Roman fort at Reculver on the Kent coast at Herne Bay. The stone implements which John found and which caused such high spirits are illustrated in Evans 1872: 536-7.
89. Evans 1943: 114–15.

90. G. A. Prestwich 1899.
91. Patton 2016: Kindle p. 12. Alice also destroyed the papers concerning his charge of plagiarism against Lyell (see Chapter 6).
92. Milne 1901: 35.
93. Fowles 1969.
94. Collins 1859.
95. Penwarden and Stanyon 2008: 49. However, there is no record in Freemason Hall of any of the principals in the 1859 time revolution being a member of a Lodge.
96. Milne 1901: 17, 95, and 97.
97. Hutchinson 1914 vol. 1: 77–8.
98. Burn (1964), named the years between 1852 and 1867, the length of a generation, as the Age of Equipoise (Cannadine 2017; Wilson 2016).
99. Thompson 2009: 115–18. The list comes from Lubbock 1887.
100. Thanks to Dan Hicks, who found this motto on the bookplates in Evans's library.
101. Smiles 1859: 145.
102. Lubbock 1894: 202; Patton 2016: Kindle p. 222.
103. Lubbock 1887; Lubbock 1889: *The Pleasures of Life*. The print run for part 1 was 272,000 and for part 2, 232,000 (Thompson 2009: Appendix 1).
104. This phrase is used by Collini (1991) to describe a nineteenth-century preoccupation.
105. Smiles 1859: 3.
106. Page 1869: 183–4.
107. Spencer 1858.
108. Martin in Eliot 1859: Kindle p. 487. *Adam Bede* impressed her peers. Dickens and Carlyle were both rapturous in their praise, the former guessing correctly it was written by a woman.
109. George Eliot 1859: Kindle p. 7724.
110. This French venture was treated with great suspicion in Britain. In its editorial of 26 May 1859 *The Times* hinted darkly that 'mischief is brewing'. British colonial interests were threatened by de Lesseps's project, which, it told its readers, 'has been all but forbidden'. Ruefully, the editor reflected that in Egypt French influence reigned supreme, even though the British had invested heavily: 'We have completed our communications with India, we have made our railway from Alexandria to Suez ... and by this time the telegraph is laid down along the whole length of the Red Sea.' And all of this thanks to a once pliant Egyptian viceroy who facilitated Britain's strategic goal of transporting troops to India and bringing goods back. Now, whined the voice of John Bull, 'Saïd Pasha [the viceroy] has surrounded himself more than ever by French advisers, and from them he learns that the duty of all noble spirits is to trample on treaties which make a province dependent on a central authority' (*The Times*, 26 May 1859, p. 8).

111. Fulgence Gigard *Le Monde Illustré*, N° 111, 28 May 1859: 347 'plus de distance
...plus de désert. Les âges mêmes n'ont plus pour lui de secrets dans leurs
abîmes'.

112. Wilson 2016: xxviii.

113. Bayly (2004: 17) identifies the development of modernity during the nineteenth
century with increasing uniformity of everyday bodily disciplines. Standardized
clothing is one of his examples and time-keeping another. The electric telegraph
made standardized time systems possible across the world.

114. Time signals were sent from Greenwich by electric telegraph from the early
1850s.

115. Correspondence in the *Athenaeum*, 16 July, 23 July, and 30 July 1859: 85, 117,
and 152.

116. https://worcester.emuseum.com/objects/30128/gentleman-in-a-railway-carriage?
ctx=c136f32b-cebe-4e26-9509-8a3675185bfd&idx=3, accessed 24 July 2020.

117. As reported in http://www.branchcollective.org/?ps_articles=krista-lysack-the-
royal-charter-storm-25-26-october-1859, accessed 24 July 2020. Many of the
bodies could only be identified by the initials monogrammed on their socks.

118. Lyell 1860: 95.

119. http://www.geomag.bgs.ac.uk/education/carrington.html, accessed 24 July
2020. These magnetograms resemble the plots that a seismograph makes of
an earthquake.

120. This war lasted from 1859 to 1861 and is also known as the Second War of
Italian Unification and the Franco-Austrian War. The Risorgimento brought
independence for Italy from the Austrian Empire and the eventual unification
of Italy's historic kingdoms.

121. Brooks 2009: 87. Losses were higher on the Austrian side, 22,000 while the
French and Sardinian forces had 17,000 casualties. By comparison Waterloo in
1815 saw 48,000 casualties, and at Gettysburg, in 1863, there were 51,000 dead,
wounded, and missing. Brooks (2009: 60) summarizes the battle as follows:
'Solferino was not the result of any grand strategic combination. The two
armies had blundered up against one another... The battle would be won by
whichever side first seized control of events.'

122. Dunant 1862: Kindle p. 821. Henri Dunant's *A Memory of Solferino* (Dunant
1862) shocked Europe. His eyewitness account of the suffering of the wounded
and the civilians led to the formation of the Red Cross in 1863 and the first
Geneva Convention in 1864. Dunant, who by then was bankrupt and living in
poverty, was the first recipient of the Nobel Peace Prize in 1901.

123. Dunant 1862: Kindle p. 1019.

124. Brooks 2009: 72.

125. Schneid 2012: 34.

126. *Pall Mall Gazette*, 28 May 1913.

127. Daguerre visited England to establish his patent in 1840; see http://www.midley.co.uk/daguerreotype/dpatent_gov_rs.htm, accessed 24 July 2020. Lyulph Avebury never saw the original but family tradition is clear that it was taken and that it showed John aged 5 standing up straight in a studio (email to author 16 February 2020).

128. Green 1953: 145. The poems were published as *Idylls of the King* in 1859 and were a huge success.

129. Hughes 2017.

130. Green 1953: 37, Letter from Dodgson to Mrs Tennyson, 4 June 1859. Dodgson had photographed Tennyson in 1857 (Taylor and Wakeling 2002: 194). The innocence of the beggar child photograph has been much debated (Douglas-Fairhurst 2017). Winchester (2011: 9) believes the costume Alice is wearing was inspired by a Tennyson poem, the meeting between the North African King Cophetua and the beggar maid Penelophon.

131. Green 1953: 146.

132. Fitzgerald 1859: stanza 60.

133. Maxwell in Dickens 1859 [2000]: 409.

134. Carroll 1865; see Gardner (1965).

135. Dickens 1859 [2000]: 385; Maxwell 1859 [2000]: 488.

136. Conlin 2014: Kindle p. 2871. The place of humans in geohistory is covered by Rudwick (2005: 275–87). Barton (2018: Kindle p. 9235) remarks on the X-Club's role in ditch-filling (see Chapter 6) by showing that scientifically there was no gap between humans and animals (Huxley 1863).

137. Mythography, which dominated intellectual endeavour for so many generations, and which George Eliot skewered in *Middlemarch* (1872) with Mr Casaubon, was beginning to wane in the early nineteenth century; Lyell's *Principles of Geology* (1830–3) was a turning point. Mythographers were interested, among many things, in the age of Adam and evidence for biblical truth. They looked for support among pagan and non-Christian beliefs and writings (Kidd 2016).

138. Trautmann 1992: 390.

139. Trautmann 1992: 384.

140. See Livingstone (2008) for a full account of the history and debates surrounding the pre-Adamites.

141. This was despite the Herculean efforts of the philosophers Kant (1724–1804) and Hegel (1770–1831), who charted the course of a progressive history that, if followed, would have done away with the ditch.

142. Drift was introduced in 1839 by Roderick Murchison to replace Diluvium, the older term for these geological sediments. He noted that whoever 'connects diluvium with the Deluge of Holy Writ must contend, that all such detritus was produced in one short period'. Geologists, however, had 'now completely

ascertained, that each region of the earth has its own superficial diluvia, produced by distinct and separate action' (Murchison 1839: 509). An unambiguous modern term was needed for this geological reality, and that was Drift, qualified by its region, as in the 'English Drift' or 'Scandinavian Drift'.

143. See Chapter 4.
144. See Chapter 6.
145. See Chapters 2 and 3.
146. See Chapter 5.
147. See Chapter 6.
148. See Chapter 6.
149. See Chapter 5.
150. See Chapter 6.
151. Evans 1875: 487.
152. See Chapter 7.
153. Babbage 1859–60: 67.

2

Discovery

The Day 27 April 1859

At seven o'clock on the morning of 27 April Prestwich and Evans were having breakfast together at Abbeville's best hotel, La Tête de Bœuf in the rue Saint-Gilles.[1] They had come to this small town in the Somme Valley to meet one of its foremost citizens. On time, the scrape of iron-shod wheels and hoofs on the cobbles announced the arrival of the child of the *ancien régime*.

Jacques Crèvecœur Boucher de Perthes quickly stepped down from his horse-drawn carriage, accompanied by the curator of the Abbeville museum, François Marcotte.[2] Seven o'clock was not early. Before meeting Joseph and John the 71-year-old had taken his daily dip in the Somme, a regime he had followed all his life.[3] His dress was unremarkable. His neatly trimmed beard in the Van Dyke style contrasted with his much younger whiskery visitors (Figure 2.1). A wig of youthful brown curls covered his baldness, and there was rouge on his cheeks.[4]

Six years earlier and he would have appeared in the uniform of Director of Customs for Abbeville, richly embroidered and with a plumed hat of which he was especially fond. But a government reorganization in 1853 abolished the position, and with the post went the uniform. Previously he would have devoted himself in the morning to what he called man-management duties[5] and in the afternoons pursued his wide-ranging musical, literary, philosophical, and antiquarian interests. In 1859 he was a full-time savant, author, and antiquary; a man of New Leisure. He was vigorous and well-off, one of Abbeville's leading lights. But the national *gloire* which he believed should be his had eluded him.[6]

Jacques was born one year before the French Revolution of 1789. At birth he had all the aristocratic advantages. Retaining that privilege during the brave new world of the Terror and the First Republic required nimble footwork on the part of his father Jules-Armand-Guillaume Boucher de Crèvecœur (1757–1844). Adroitly, he distanced the family from the dangerous values of the *ancien régime*. He moved the noble house of Crèvecœur to

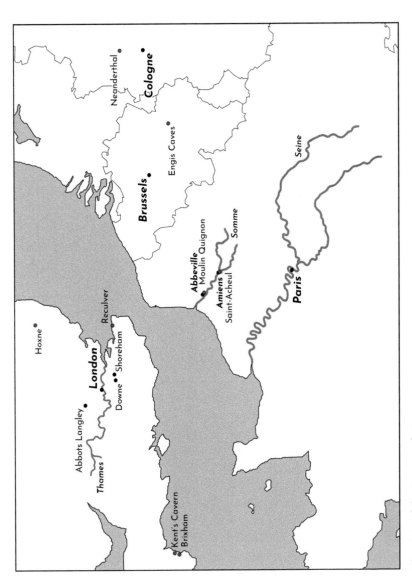

The sites of the time revolution

Source: Kaylea Raczkowski-Wood

Figure 2.1 The founding father of deep history. A bust of Jacques Boucher de Perthes by Emmanuel Fontaine (1856–1935) in the Musée d'Archéologie Nationale at Saint-Germain-en-Laye. Posthumously commissioned to accompany his collections.
Source: © RMN-Grand Palais/Philippe Fuzeau.

Abbeville in 1792 and took up the government position that would be passed on merit to his son in 1825. He rarely used his title, being known instead as Jules Boucher, which suited the *égalité* of the First Republic and the First Empire which followed it. Jules prospered irrespective of how France was governed, and in 1803 the Bouchers had bought the imposing Hôtel de Chépy, an eighteenth-century mansion on rue des Minimes in the centre of town.

Jacques continued the tradition of playing down the noble Crèvecœur. But he embellished plain Jack Butcher by tacking on his mother's maiden name, Boucher de Perthes. Scratch them a little, however, and they remained Crèvecœurs, Heartbreakers.

Jacques was the second of six boys. He had one younger sister Aglaé. Only one of his siblings, Étienne, outlived him, and Jacques never married.[7] In his late teens he cut a lively figure in the Mediterranean city of Genoa, a far cry from northern Abbeville. There were rumours of a liaison with Napoleon's sister, Pauline Bonaparte, a Borghese princess.[8] He once had to hide, on her orders, for four hours in a wardrobe. By the time she released him, he was nearly asphyxiated, so perhaps a further rumour that their tryst was never consummated is unsurprising.[9] Jacques was imaginative, too imaginative at times, and his eight-volume autobiography *Sous dix rois* (Under Ten Kings) cannot always be trusted.[10] He could be a fantasist, inventing correspondence that inflated his looks, achievements, and pedigree; a good example of the last: adopting the name Boucher *de Perthes* to ring distant bells with Joan of Arc.[11]

Those who have read his massive literary and antiquarian output usually declare themselves impressed, frustrated, and underwhelmed in equal measure.[12] He published forty-nine books in sixty-nine volumes. It is quite likely that, if stacked, like the geological layers in the quarries that finally won him the national glory he craved, they would have topped his height of 5 feet 6 inches. This tottering *œuvre* covered most subjects and included a novel for good measure, *Emma, Some Letters from a Woman* (1852). Its heroine is a beautiful English aristocrat and homicidal monomaniac Lady Emma de North***. She has a devoted French suitor Jules de P** whom she repeatedly attacks and who constantly forgives her. If the fictional suitor is his father Jules, or perhaps his older brother, another Jules, then the format of the book becomes disturbing, as much of it is written in the first person. But there are enough clues that in this case Jules de P** was Jacques himself: constant references to his striking good looks and sobriety. Jules de P**, like Jacques, only drinks water.[13] And Jacques was the only one in his family to add de Perthes to his name.

The novel presents women like Emma as innocent monsters set apart by their nature.[14] And in another mixing of fictional and personal histories, Emma's homicidal nature is shaped by the political upheavals following the fall of the First Empire in 1814. These touched Jacques also and led to a dip in his career. Demoted from his role as *sous-chef du personnel* (man-management again) in Paris, he was sent as Inspector of Customs to the small Mediterranean port of La Ciotat. Here he languished until 1816, when a posting to Brittany revived his career.

From the evidence of *Emma* it seems that women unsettled Jacques. After all Jules de P** meets his end, stabbed in the back by his lover with a dagger bought from an antiquary. Does the novel also give a clue about the character of its thinly veiled author as her madness contaminates the novel's strangely feminized Jules de P**?[15] Was Jacques, like Emma de North***, shaped by family and political events that irredeemably bruised his ego? Or was he, as is suggested by his stack of books, prone to a hyperactive condition? Nature, nurture, or a bit of both? The archaeologist James Sackett sums him up best: possessed of a butterfly mind, Jacques was no genius, 'but he may have had a greater variety of talents than can comfortably fit into any one person'.[16] The eminent British geologist, and namer of the Drift, Sir Roderick Impey Murchison (1792–1871), met him in Abbeville in 1861. He declared, superciliously, that 'he was one of the most completely original French characters I have known for many a long day'. He recognized Jacques's ability to enthuse others but regarded him as 'an old fashioned "gobemouche", very credulous, and easily imposed upon'.[17] Prestwich, writing after his earlier visit, delivered a kinder judgement: Monsieur Boucher de Perthes is 'an antiquary distinguished by his varied researches, his large and valuable collections, and by an indefatigable zeal and perseverance'.[18] Zeal was a favourite Prestwich word and perseverance the mantra of Smiles's *Self-Help*.

What we can say is that the Bouchers prospered by not being the Crèvecœurs. Jules's and then Jacques's long tenure as the Directors of Customs for Abbeville, from 1792 to 1853, was marked by good fortune. They skilfully dealt with the impact on French trade of British blockades and sanctions when the two countries were at war. Then in 1835 the Canal Maritime was completed joining the Somme to the Channel port of St-Valéry. Freight was unloaded at the elbow of the Somme, the Quai de la Pointe, as it flowed west out of Abbeville and from there dispersed throughout the canal system of France. Location captures flows, and flows capture revenues as well as history.

Jacques survived three kings, two republics, and two emperors, Napoleon Bonaparte and Napoleon III (the 4-year-old Napoleon II, who reigned for a month does not really count but was one of Jacques's ten kings). If he had been born an Englishman, he would have had the dynastic stability of three Hanoverian kings, one of whom, George III, became deranged, and one queen, Victoria, who was rescued for a while from a peevish mind by Albert, her sharp-witted but short-lived German prince consort.[19]

Jacques was 27 when the long eighteenth century came to an end on the battlefield of Waterloo. He avoided military service and was above all a skilled bureaucrat who thought the route to improvement lay in reforming institutions rather than trying to improve human nature. He wrote extensively about the values of *liberté*, *fraternité*, and especially *égalité*—but was never tempted by political office or the redistribution of wealth. In his political tracts he championed equal rights for women.[20] He may have been born in the *ancien régime* but he had adapted, like his father, to the new political environment. The Crèvecœurs' ability to change followed the path proposed for biology by Jean-Baptiste Lamarck (1744–1829) rather than Darwin, passing on characteristics acquired during one lifetime to the next generation. In Jacques's case his memes rather than his genes have long outlived him. More than anything he liked to think outside the box from the security of the Hôtel de Chépy.

It is too easy to use Jacques's words against him and brand him as a dilettante. He desired to be taken seriously and craved election to the Académie des Sciences in Paris, or at least to follow in his father's footsteps as a corresponding member. Neither happened. He was vain, bordering on the narcissistic, but he was also diligent, as his work in the Customs shows. His great quality was that once he had adopted a cause, such as demonstrating an antediluvian human antiquity, he persevered. Joan Evans dismissed him as an idle man (he was retired) and 'a good walker'.[21] Grace McCall, who met him, was far more perceptive. She described the Hôtel de Chépy as a roomy old house which did not have a single habitable or comfortable-looking room in it, 'and must have been a dreary abode for any other than its owner'.[22] Jacques was isolated, living on lonely street at Heartbreak Hôtel. But he was not a frivolous *flâneur*.

By comparison John and Joseph were sedate, reflecting their ordinary 40-year-old monarch swathed in her layers of crinoline. What you see is what you get with these two. They were never tempted to write fiction, quite the opposite. Neither of them aspired to an autobiography, let alone ten volumes of self-praise. They worked hard in their family-based businesses. Joseph must have envied Boucher de Perthes's position, where he could devote himself full-time to the pursuit of his scientific passions. He had to fit his into and around business trips. If John had possessed the capacity for self-reflection, the lack of which his youngest daughter Joan bemoaned, then he might have recognized common ground in the manic drive and quest for recognition. But the lack of sustained focus would have driven him to despair. Heartbreak Jacques was a starter of ideas. The highbrow Joseph had the means to put flesh on them, while steadfast John was a finisher. This combination of turbulence, momentum, and persistence that came together on 27 April was about to make deep history possible.[23]

Pits in the Drift

It was the first time the three men had met.[24] They conversed in French. Prestwich went to school in France and his wine business would be impossible without fluency in the language.[25] Evans was a skilled linguist.[26]

They were now in the carriage and eager to begin the day's business with a tour of the working quarry pits of Abbeville (Figure 2.2). Their first call was to the north of the town through the Porte Mercadé that guarded its eastern approaches. Their destination was the nearby village of Menchecourt. Then they doubled back and crossed the Somme to reach the pits on the chalk hills above Rouvroy about three miles to the west. Then back through the town again to visit the gravel beds to the south-east at Moulin Quignon on the road to Amiens.

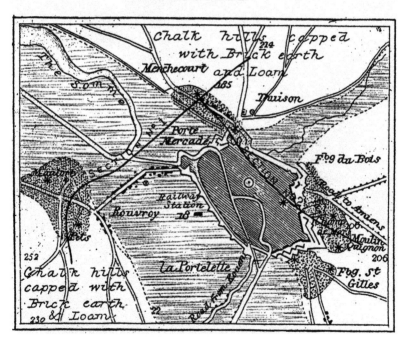

Figure 2.2 Prestwich's 1860 map of the gravel pits at Abbeville drawn by his sister Civil. The two sections that Prestwich drew across the town are shown with black lines, and the asterisks identify the major gravel pits.
Source: Prestwich 1860.

The deposits dug for a living by the *terrassiers*, quarrymen, at Abbeville were known to English geologists as the Drift: sands, clays, gravels and brickearths.[27] As Evans put it, the deposits of drift are spread out over the older rocks such as the chalk of southern England and northern France 'by the driving actions of currents of water, whether salt or fresh, or by the drifting action of ice'.[28]

Their quest that morning was highly specific. They were looking for a distinctive type of stone, humanly shaped into a tool. Only Jacques had seen such a remarkable thing. But they wanted more than that. They wanted to witness it being plucked from within the gravels of the Drift where it had lain for a very long time alongside the bones of extinct animals.

These tools had been known to Jacques and a small group of like-minded Abbevillois for more than twenty years.[29] His curiosity was stimulated in 1837 by the work of a young physician Casimir Picard (1805–41) who was a pioneering antiquary on the Somme. Picard died young in 1841 but not before he had published in the *Mémoires de la Société Royale d'Émulation d'Abbeville* in 1834–5 a ground-breaking paper that identified stone implements coming from geological deposits.[30] Moreover, in the same volume of the *Mémoires* there was a paper by the geologist François-Prosper Ravin (1795–1849)[31] who noted a difference between the upper peats and river silts and the lower, and hence older, sands, clays and gravels. As archaeologist Donald Grayson points out it was not difficult to put the two papers together and reach the obvious conclusion, that at Abbeville there were at least two periods with two sets of distinctive artefacts.[32] The person who made that connection and pursued it was Boucher de Perthes. The possibility of finding evidence of humans before the biblical flood combined for him the fields of philosophy, archaeology, and geology.

As a geologist Jacques was a fully paid-up catastrophist. His antediluvian animals and the humans who made those stone tools were wiped out by a great flood.[33] This was not Noah's flood, which he also believed in, and unlike the deluge in Genesis nothing escaped this older cataclysm. Antediluvian humans and their stone tools had to be created all over again. The possibility of continuity, the stuff of Darwinian evolution, from a remote ancestor to a living population did not cross Jacques's mind, and in that respect, he represents in 1859 an antiquated branch of geological thinking.[34] His visitors that April day, while never keen evolutionists were nonetheless at the forefront of the modern geology.

Menchecourt was Boucher de Perthes's first artefact locality. For a long time its pits had produced unmissable large bones of woolly mammoth, woolly rhino, bison, lion and other species from the extinct bestiary of the ice age. Many of these were sent to Paris to appease the hunger for fossils and several of them landed on the desk of the foremost naturalist and classifier of the day, Georges Cuvier (1769–1832). But what Picard and then Boucher de Perthes were suggesting went much further. Their claim was that some of the stones found with these bones were ancient tools. Such a claim was a stretch not just because humans were only as old as the Bible allowed them to be but because nobody had demonstrated that these curiously shaped stones were in fact man-made artefacts. They were in unknown territory.

Jacques found his first unpolished stone artefact at Menchecourt in 1841.[35] He had instructed the *terrassiers* to dig out a mixture of sand and flint from the bottom of the pit and place it on a canvas groundsheet so that he could sort through the material. It was in this fashion that he found an unpolished axe, 22 cm in length. It became artefact Number 2 in his sketches of antediluvian axes in his monumental book.[36] But when, a hundred years later, Jacques's spidery drawing was matched by archaeologist Leon Aufrère to the original specimen that still existed in the Musée Boucher de Perthes it was obviously a Neolithic axe and much later in age.[37] The same is the case for his Number 3 on the same plate of drawings although it was found at a depth of seven meters by a local man, François Corbillon who according to Jacques, with a typical flourish, was known as the 'lark'.[38]

There were no such question marks over his next discovery at the Hôpital des Pauvres in the town. In May 1844 the Corps of Engineers began to dig a large, deep trench behind the Hôpital and between the Fairground[39] and rue Millevoye.[40] The work lasted all summer and into the autumn. It presented Boucher de Perthes and the geologists of Abbeville with the opportunity to trace over some distance and at different depths the implement bearing sands, clays and gravels of the deep Drift. This was a picture window rather than the usual peep-hole into the deep past. Several sections were drawn along its length by geologist François-Prosper Ravin and they began to tell a consistent story; between about three and five meters (10–16 feet) below the surface they could trace a lower layer of sands and clays within the gravels laid down by the Somme. These pockets of soft sediment, as well as the gravels, regularly contained the bones of extinct animals and stone implements similar to those found at Menchecourt. Unintentionally the Army Engineers had provided the circumstances for the geologists of Abbeville to carry out a repeatable experiment. The results did not disappoint.

One of the geological sections from this trench near the Hôpital formed a key part of Boucher de Perthes's case (Figure 2.3). It appears in the first of

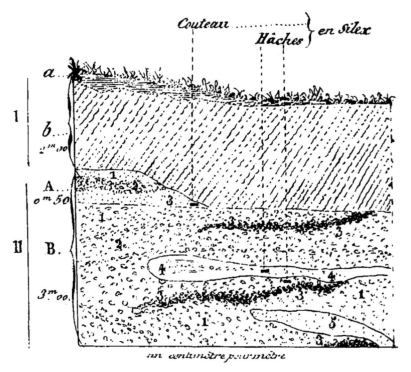

Figure 2.3 Hôpital section in Abbeville as seen by Boucher de Perthes in the summer of 1844. The sections were cut by the Corps of Engineers during work on Abbeville's defences and were no longer open in 1859 for Evans and Prestwich to inspect. The positions of the axe and the knife are indicated by the dotted lines.

Source: Boucher de Perthes 1847.

his three-volume classic, *Antiquités Celtiques et antédiluviennes: mémoire sur l'industrie primitive et les arts à leur origine*; published in Paris in 1849 although the title page has the date 1847.[41]

The section in the drawing is just over three metres (10 feet) in depth. It starts with a large slice of modern topsoil and rubble from earlier buildings on the site. This is the upper layer from the surface of history. Nothing for us here. It is the lower level (II) where things become interesting with clays and sands in the first bed (IIA) and gravels in the second (IIB). The prize lies in the dotted lines that point to a stone knife (*couteau*) at the base of IIA3 and stone axes (*haches*) in a vein of sand within the gravels (IIB4). These levels also produced extinct animals. Their presence led to Jacques's great claim. These animals did not roam their chilly world alone. They were accompanied by humans who did not leave their skeletons behind. Instead they left their handiwork. This was the claim that John and Joseph had come to inspect.

The title of Jacques's 1847 book says a great deal. It refers to two periods of time; a recent one from the shallows of history, the Celts, and an antediluvian era, before the biblical flood: the window into a new, deep history. This was Jacques's great insight; that there were two distinct time periods, differentiated by the contents, antiquities and fossil bones, of their geological layers. By setting them in opposition to each other he promoted the claims of the older. To imagine a deep history of geology and stone implements he needed to bounce it against the familiar history from a shallower time-depth. In the upper levels wild and domestic animals like red deer and sheep were found alongside polished stones made to be hafted as axes. Something Picard had demonstrated. Beneath them in the gravels and the sands were extinct woolly mammoths, giant lions and unpolished handheld axes. He needed to separate out these two levels. But at the same time in order to make his case that humans lived in the older stratum he had to drop a ladder down into the depths so that sceptics could take the journey with him. The rungs in that ladder went from the recent to the remote, Celtic to Antediluvian, polished to rough axe.

Langues de chat

A visit to a quarry to find stone artefacts is the triumph of hope over expectation. Most of the time you find nothing and all you get are extremely muddy boots. In winter quarries can be some of the bleakest places on earth. As archaeological places go they lack enchantment.

Nowadays quarries can be dangerous places with huge machinery and faces which can collapse. Hard hats, hi-vis jackets and proper boots are mandatory as are risk assessments, insurance and permissions. Not so in 1859. These pits were dug by hand in steps, the sands and gravels moved by wheelbarrow and cart. The most sophisticated piece of equipment was a sieve to clean and sort the gravel into sizes so that it could be used as aggregate. The pits were not owned as is the case today by large, often multinational companies, but by small-holders and worked by families who supplemented their income by digging aggregate when it was needed. The pits were named after their owners Dufour and Deliquière at Menchecourt or T. and F. Fréville on the Cagny Road at Amiens.[42] Pits were highly sensitive to the demand for ballast and building and Abbeville had seen plenty of that with the arrival of the railway in 1847 and improvements to the roads. Pits provided seasonal work; the lower part of Dufour's pit was worked in winter when there was a dearth of agricultural work. One of the men who moved the sand from Menchecourt was the local gravedigger, François Hautefeuille.[43] These quarries were marginal in every sense; to the town, to incomes and to the economy of the local area. But that does not mean they are unimportant archives for human history.

As the morning progressed Prestwich began to put what he saw on to a plan. He used the pits and the heights of their deposits to draw a cross section across the valley from Menchecourt to Rouvroy and then another along the line of the river through the town from Menchecourt to Moulin Quignon. Height was the key for Prestwich; measured using an aneroid barometer.[44] The hills around Abbeville rise to between 185 feet and 205 feet. The base of Dufour's pit at Menchecourt is 42 feet above mean sea level at the Somme's port, St Valéry. Within this valley his plan identified three levels of widespread Drift. The topmost were silts, peats and some gravel found mostly at the flood level of the present Somme. These were very recent. Then came brickearths and beds of angular, rather than river-rolled, gravel. The third and lowest level was made up of sands and gravels and it was these that contained bones of extinct animals and stone tools. His cross section is a feat of geological extrapolation. From small windows into the deep past, such as the Hôpital (Figure 2.3), he inferred continuous beds of similar sediments stretching across the valley. He confirmed, if it needed to be, the observations of Ravin and Boucher de Perthes and the Parisian geologist Charles-Joseph Buteux (1794–1876) who had made a special study of the Somme in 1857.[45] What gave Joseph's great slice through the Somme its authority was being able to tie it in to the admirable contour maps made available to him by the Département de la Guerre.[46] One site, however, stood out as different. Moulin Quignon lay at a higher elevation, 106 feet, and its gravels were shown by Prestwich as a capping on the hill rather than part of the lower beds of Drift.[47]

The *terrassiers* had known Jacques for many years. In his book he mentions their honesty and integrity. He knew them by name as any serious collector must. An insight into this special relationship was provided by Worthington G Smith (1835–1917), a curmudgeonly but inspired collector of worked flints principally in Bedfordshire. Looking back in 1888 on his experiences he had this advice:

> As a rule, gravel diggers don't like 'mashers' [dandies], or gentlemen with lavender kid gloves, they are not very fond of clergymen, and they object to thoroughbred idlers. To make a good beginning with strange diggers a man should not be too well dressed; he should jump gaily into the pit, and at once begin conversing in a friendly and familiar fashion with the men. A little harmless slang or good humoured banter sometimes goes a long way.[48]

Once instructed by Jacques to look for these stones the workmen obliged the eccentric gentleman who paid well. They had their own name for these stone axes, *langues de chat* (cats' tongues), a popular flat crunchy French biscuit, straight sided with rounded ends and a convex cross section (Figure 2.4).

The *terrassiers* became proficient in finding these oval and pointed cats' tongues and at Menchecourt they had the recent finds ready to show the

Figure 2.4 The distinctive shape of these biscuits gave their name to the stone implements the quarrymen were finding in considerable numbers at Abbeville and Amiens. Cats' tongues also come in milk chocolate, *Katzenzungen*, and are extremely addictive. Any number of recipes for *langues de chat* can be found on the Internet.

Source: Author's photo.

English visitors. This was the first time John encountered such an unpolished implement, what Boucher de Perthes called a *hache ébauchée*, rough axe. Nothing like it was known in England. By contrast polished examples were common in both Britain and France and across Europe. Jacques was asking John to accept that polishing distinguished a recent from an ancient past.[49] Polishing an axe by hours of constant grinding and smoothing was so obviously an indication of human handiwork. What had to be accepted with the unpolished examples was that the pattern of flaking, removing chips from a nodule of flint by knapping it with another stone, were equally diagnostic of human action. But what concerned John most was the possibility of forgery. After all, the *terrassiers* were paid for their discoveries. Boucher de Perthes was not always present when they found the implements. Evans was well aware of the thriving cottage industry in forged coins and metalwork planted in the Thames foreshore and offered for sale to collectors like himself by Victorian mudlarks such as Billy Smith and Charlie Eaton.[50] A few years later and John had to be on his guard for the prodigious output of 'Snake Willy' (William Smith) and the notorious 'Flint Jack' (Edward Simpson) whose efforts made sure that Yorkshire led the way for forgeries of flint weapons of all descriptions.[51]

As they ticked off the pits on their list they were accumulating these curious artefacts, animal bones and jars of shells from the sands and gravels of the Drift. The shells, marine and freshwater, were useful for indicating changes in tidal reach as well as climatic conditions and for slotting deposits into Charles Lyell's scheme, devised in 1839, for the recent geological epochs that he called Pleistocene and Holocene.[52] Prestwich also measured and sketched sections at Menchecourt, Moulin Quignon and the Porte Mercadé which afterwards were drawn to publication standard by his sister and included in his paper to the Royal Society in London.

Disappointment, however, clouded Joseph's day because up until now 'the search for flint-implements was not so successful as I could have wished'.[53] There were moments of excitement. The Menchecourt *terrassiers* held up 'enveloped in the sandy matrix (a portion of which was agglutinated on the specimens), three thin sharp-edged flint-flakes, 3 to 6 inches long'.[54] 'Flint-knives' pronounced Jacques without hesitation, and Joseph and John followed his logic because they differed from flint splinters produced naturally by stones bashed together in a river or chopped by an iron shovel. Under Jacques's guidance they saw that one of these knives was struck with another stone in an attempt to remove an unnecessary edge so that it became thinner, more knife-shaped.

Ever the stickler for decisive proof Joseph stood supervising the *terrassiers* digging the gravel.[55] What he needed was a fresh, vertical face to the quarry wall; not only to see how the layers related to each other but also to have the best chance of finding one of these 'so-called flint hatchets' *in situ* in either the sands or the gravels of the lower Drift. But he was ready to be convinced. He questioned Boucher de Perthes, and received confirmation from the *terrassiers*, that at Menchecourt the *haches* are most common in the lower part of the sand and in the gravel.[56] There was he believed enough 'moral and collateral testimony' to agree with Boucher de Perthes's view that these stones were indeed stone implements and that their age was antediluvian, therefore very old.

So a good, but not conclusive morning's work ended. They got back into the carriage with their jars, boxes, drawings and flints and returned to the Hôtel de Chépy for lunch.

Déjeuner à la fourchette

Once through the impressive front door of Boucher de Perthes's home, they were surrounded by the collections that his father began and he expanded. Evans recalled in a letter he wrote to his fiancée Fanny Phelps on Sunday 1 May that the house 'is a complete museum from top to bottom, full of paintings, old carvings, pottery etc. and a wonderful collection of flint axes and implements found among the beds of gravel and evidently deposited at the same time with them'.[57] Murchison, during his later visit, was less impressed: 'His house and furniture are indicative of his mind,' he snarked. 'Every sort of villainous copy put out as originals, and curiosities of all sorts.'[58]

A photograph of the salon de musique taken in 1894 shows floor-to-ceiling glass-fronted cases and flat-topped vitrines filling every inch of the room.[59] The doorway is flanked by marble busts, while the antlers of ice age deer hanging high on the walls make it feel like a Stone Age hunting lodge. Joseph, unlike Sir Roderick, was 'much struck with the extent and beauty of M Boucher de Perthes' collection'.[60] There was a great ebony crucifix, Italian faience, medieval icons in ivory, wood, and copper, ceramic figures from the fifteenth and sixteenth centuries, medals and medallions, heavy furniture, and many drawings and paintings. Among them were two portraits of Jacques himself: one as Coco, the precocious 10-year-old child,[61] and another as the handsome 43-year-old Jacques in his Customs uniform.

After that the portraits stopped. Once he reached middle age and became a wig-wearer, he no longer sat for portraits, and there are no photographs.[62] There is, however, a posthumous bust of the elderly Boucher de Perthes by Emmanuel Fontaine (Figure 2.1) and an effigy in bronze on his very public tomb.[63]

In his museum the stone axes were interspersed with cases of the bones of animals from the antediluvian levels, all of them extinct. Put the two together and we have in John's excited words to the woman he would soon marry, 'the remains of a race of men who existed at the time when the deluge or whatever was the origin of those gravel beds took place'.

But first there was a formality to be observed, a visitors' book to sign.[64] This was brand new, so there was no record of last year's visit by Dr Hugh Falconer and his niece Mrs Grace McCall, recently widowed at the age of 23. They visited Jacques on a bitterly cold 1 November en route to Sicily to escape the English winter. They were there because of Boucher de Perthes's 1847 book and wanted to see his collections of flint tools at first hand.[65]

Grace later wrote an account of the meeting.[66] They had arrived a day late, dropped their bags at the Tête de Bœuf, and hurried on to the large old building set behind its iron railings in the rue Minimes, only to find it shuttered and with a carriage filled with luggage waiting outside. 'Five minutes later', she wrote in 1895, 'and another hand would have had to chronicle the first recognition by English men of science of the old flint implements of the Somme Valley.'[67] Jacques had given up on them and was returning to the country. But everything soon changed, and their host made them very welcome. As Uncle Hugh wrote, he was 'a very courteous elderly French gentleman, the head of an old and affluent family'.[68] Jacques insisted on showing them the wonders of his medieval collection, and the tour began in his overcrowded study. Eventually they were shown to the geological gallery where the flints were kept. 'The collection was a magnificent one,' Grace recalled 'and our host was almost breathless with excitement in detailing the circumstances in which each specimen had been found.' It soon became a case of information overload as they spent the rest of what she called a 'memorable' day with the flints. Uncle Hugh was deeply fond of his niece for many reasons but none more so that day because she was a trouper. There is a tradition with breakthrough discoveries that there is no gain without pain, and Grace could still remember thirty-seven years later that the gallery 'was like an ice house, there was no fire, and the very handling of the flints was freezing work', so much so 'that it nearly finished the unfortunate secretary' herself.[69]

That evening as they thawed out before a roaring fire in the Tête de Bœuf, Uncle Hugh sat down and wrote to Prestwich strongly recommending that he find the time to visit. He was convinced by the evidence and saw that it could settle, finally, the speculations 'regarding the remote antiquity of these industrial objects, and their association with animals now extinct'. And then in a flourish intended to flatter he declared, 'You are the only English geologist I know of who would go into the subject *con amore*'—a prescient phrase in view of the presence of Grace, his future wife. What concerned Falconer was that English geologists were underestimating the importance of the Drift for human antiquity. Neither were these solely English blinkers. The same lack of vision applied to French geology; and why? Because 'Abbeville is an out-of-the way place, very little visited, and the French savants who meet him in Paris laugh at Monsieur Boucher de Perthes and his researchers.'[70] And not just the French. The savant of savants Charles Darwin had pronounced on his catastrophist theories 'that the whole was rubbish'.[71]

The Hôtel de Chépy was also the museum and meeting place for Abbeville's learned society, the *Société Impériale d'Émulation* (it had to replace its *Royale* status with the arrival of the Second Empire in 1852). Societies such as these spanned interests as wide as Jacques's and were a source of civic identity and pride.[72] Unsurprisingly, in 1859 it was an all-male club. Picard had been a member before his early death, and other members included the geologist Ravin. Jacques had served as president of the society since 1830, when he breathed life back into it. What John and Joseph were being asked to accept touched on the reputation of all Abbevillois who supported their president.

There was much to see. Aided by Marcotte, the curator of the Museum, they were shown cases of flint tools. Jacques flicked through his book to the description of the implements from the trench behind the Hôpital. He opened a case and pulled out two stone implements, an axe [*hache*] and a knife [*couteau*]. These came to light in 1844 on 23 and 24 July respectively and at a depth of between 4.5 and 5 metres.[73] Because of their greater depth, he regarded them as the oldest implements yet found.[74] He had them sketched and described, along with many others (Figure 2.5).

As to the association of these stones with extinct animals, Jacques asked if the Englishmen had not seen that morning such occurrences for themselves and spoken to the quarrymen who confirmed it. He turned to the well-thumbed page 263 of his book. On 7 August 1844 he arranged for the two *terrassiers* involved, Philippe Courbet and Louis-François Forteguerre, to

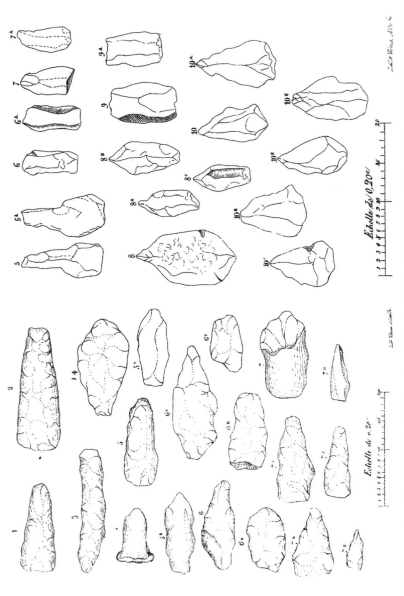

Figure 2.5 The axe found in 1844 is No.6 in Boucher de Perthes's Plate XVII (Left) and the knife is No.8 in Plate XXVIII from his 1847 book (Right). The sketches are of poor quality, but he added scales, and descriptions in the text.

Source: Boucher de Perthes 1847.

sign a *procès verbal* (a form of affidavit) as to the veracity of their discovery.[75] Courbet signed his name and Forteguerre made his mark. The document was independently witnessed. He handed them the stone tools the men found and which surely, he pleaded, any rational person would regard as proof that humans existed with the mammoths and other extinct animals.

Jacques was now pushing at an open door. Joseph and John had seen enough to concur with Falconer's good opinion of their congenial host. His claim for an *industrie primitive*, those unpolished stone axes, was no longer outrageous, because it was amenable to a repeatable scientific experiment by finding more axes embedded in deposits of Drift.[76] The weight of evidence had tipped the balance. Prestwich continued to ask more questions, this time about the staining of the flints. He had noticed that those from Menchecourt were white and bright, and where some of the matrix in which they were found still stuck to them, that too had the same lustre. By contrast the artefacts from Moulin Quignon were dull yellow and brown. Such staining of artefacts in sands and gravels is commonplace as are the well-rolled edges when they have been part of a river's load.[77] Both observations were an argument against forgery.

But, as often with Jacques, there was a sting in the tail; and today that barb was the last part of the title of his ground-breaking book: '... *et les arts à leur origine*'. After the stone tools came the flint animals. He devoted the last two chapters of his book to stone figures and symbols, first of the Celtic period and then from the much older antediluvian Drift. But what he had done was to sort natural pieces of gravel into similar, familiar shapes. One plate among many will suffice (Figure 2.6). These, he claimed, were the heads of roe deer, collected by ancient man.[78] They were the number 8s in his classification. Number 9s were the heads of horses, and so on through a total of twenty-four categories. Jacques was convinced that he had uncovered not only workshops for stone tools but also studios for the production of antediluvian art.

Joseph and John swallowed hard. These were nothing, figures in the flames, camels in the clouds. The idea of the *objet trouvé* that Marcel Duchamp will use to challenge the very idea of art is fifty years in the future. And that was definitely not Jacques's intention. Prestwich was diplomatic but firm. There were many forms of flints among which he 'failed to see traces of design or work, and which he should only consider as accidental'.[79] The laughter in Paris was partly at the exaggerated claims for these natural objects.

Drawer after drawer was filled with these unmodified stones. And it made no difference that they came, in Boucher de Perthes's opinion, from both the

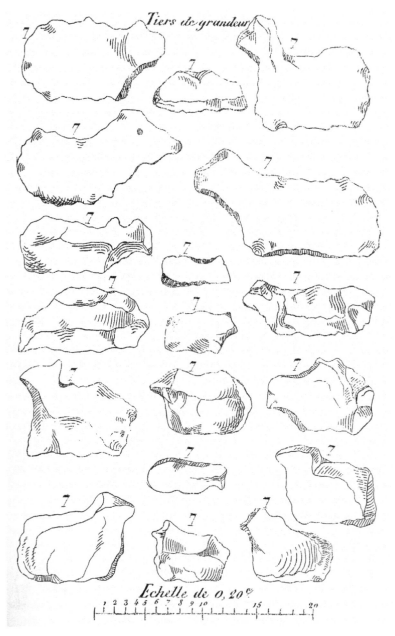

Figure 2.6 Stones that Boucher de Perthes believed were chosen by antediluvian Man because they resembled the heads of roe deer.

Source: Boucher de Perthes 1847.

Celtic and antediluvian periods. This was one instance where playing one period off against another did not work. Nonsense meant nonsense.

Jacques is not alone in getting one thing, primitive technology, spectacularly right and then another, figured stones and the origins of art, horribly wrong. There is a facet in intellectual pioneers like Jacques that confounds. Isaac Newton is the best known for his mix of fundamental physics with the study of the occult and the practice of alchemy.[80] Darwin's peer as a naturalist, Alfred Wallace, extolled the virtues of phrenology to discern human character and excoriated vaccination as bad science.[81] One of the leading British antiquaries of the eighteenth century, William Stukeley, rescued the Wiltshire monuments of Stonehenge and Avebury from obscurity. He set in train a way of seeing landscapes as a palimpsest of historical periods: one that incorporated monuments which had no history other than folklore passed on in oral tradition. He drew their first plans and, given the subsequent ill-treatment some of the monuments received, these are invaluable today. But Stukeley also created the white-robed Druids.[82] This order, which is not even 250 years old,[83] has connected itself to Stonehenge, which is 6,000 years old. History is full of such invented traditions that collapse time by folding a much deeper past into the contemporary world.

The idea that a deep history monument such as Stonehenge was built by Stukeley's Druids, is now so firmly lodged as a quasi-historical 'fact', along with cavemen living with dinosaurs and Vikings with horned helmets, it is impossible to budge. So too with these figured stone animals which start with Boucher de Perthes and still have avid collectors today. These objects belong with psychologists rather than archaeologists. Similar things will sweep up Joseph in his old age, when he became entangled in the sticky mess known as the Eolith (Dawn stones) controversy.[84] Thirty years earlier in 1859 he was on top of his game and not standing for any nonsense.

Food for thought, but they were saved from indigestion by the announcement of lunch. And it was not just 'lunch' but, as John wrote to Fanny, 'a sumptuous *déjeuner à la fourchette*', the full cold collation with wine, served buffet-style. This presented a slight paradox, because the generous Boucher de Perthes was a teetotaller and ascetic. But he knew how to lay on a spread that thawed the glacial in Sir Roderick Murchison. 'A right hospitable and good French gentleman, [who] gives you good cheer and capital wines', the Scottish-born geologist wrote of his host, whom a page earlier in his journal he described as over-credulous, a *gobemouche*.[85]

Over the knives and forks the discussion got back to *haches* and *couteaux*, the delights of antiquarian pursuits, and geological research.

Travelling Companions

A herald of nineteenth-century progress, unheard by those enjoying Jacques's hospitality, knocked on the door of the Hôtel de Chépy and handed in a telegram.[86] The envelope with its slip of paper was carried into the dining room and presented to Joseph. Its message was urgent. The proof he was looking for had been found in Amiens. They should come as quickly as possible to a pit outside the city at Saint-Acheul. The sender will meet them at the station and take them there.

Boucher de Perthes insisted they lose no time and the Englishmen pushed back their chairs and started making their farewells. Within minutes they were in the carriage and heading for the railway station to the west of town. A southbound train awaited.

Telegraphs and railways went together. Electrical impulses extended an individual's reach down the line, just as the railway tracks they marched beside conjured a new sense of connection between people, towns, and landscape. The nineteenth century saw those tracks as a journey of progress measured by an insistent new tempo; the world now divided into those on board and those left behind. This mood was captured in 1857 by the travelling companions Charles Dickens and Wilkie Collins:

> It was like all other expresses, as every express is and must be. It bore through the harvest country, a smell like a large washing-day, and a sharp issue of steam as from a huge brazen tea-urn. The greatest power in nature and art combined, it yet glided over dangerous heights in the sight of people looking up from fields and roads, as smoothly and unreally as a light miniature plaything.[87]

The train to Amiens set off, pulling a string of carriages which looked as if they had been commandeered from redundant stagecoaches. Once aboard, John and Joseph were in familiar surroundings. They were inveterate railway users for both business and geological excursions. They settled back and started to reflect on their morning.

As the train rattled along, stopping at Pont Rémy, Longpré, and Hangest, they had time (the stopping train took an hour and twenty minutes[88]) to digest their lunch and their impressions of Boucher de Perthes. They might have laughed about his stone animals but were suitably impressed by the geological deposits of Drift he had found. John no doubt enthused about the stone tools, the *haches*, that he had seen for the first time that morning and how it might be possible to copy their manufacture. After all, gun flints were a component for muskets and skilled stone knappers turned them out by the thousand at Brandon in Suffolk.[89] Those skills could be useful for a future experiment. But above all Boucher de Perthes's collection had raised a conundrum. Flint was commonplace in England. So why were Jacques's roughly made axes (*haches antédiluviennes*) found along the Somme, but nowhere else, and to their knowledge not in the English River Drift? Moreover, they were unlike the stone implements that Prestwich had seen at Brixham and which Evans had yet to handle. Gradually they settled back and took in the views of the valley.

Railway travel in 1859 encapsulated enchantment and disenchantment, the human ingredients through which surroundings are experienced and histories flavoured. A painting in 1862 by British artist Augustus Egg of two female travelling companions sums up the former (Figure 2.7). Billowing, they fill the space in a sumptuous fashion, while out of the window is a glimpse of the exotic Mediterranean coast at Menton. The swinging tassel at the window conveys the movement of the train. The women are distinguished by two objects: a book and a basket of fruit, symbols for some of industry and idleness but also containers of enchantment and imagination. Otherwise, the women are symmetrically similar. How different they are to a contemporary painting by Honoré Daumier of the *Third-Class Carriage*. His disenchanting image shows the impact of industrialization on lives and imagination—no views at all from these windows, no books to read—and the gulf between first and third class in the country of *égalité*. Progress seen through the nineteenth-century eyes of the two women, one breast-feeding her child and the other carrying a working rather than decorative basket, is imaginatively impoverished, modernity's hollow dream.

The train to Amiens had now passed through Picquigny and Ailly. Put two friends like Evans and Prestwich on a mission and it is difficult to escape comparisons with two other railway habitués, Holmes and Watson, created by one of 1859's birthday boys, Arthur Conan-Doyle born on 22 May. There is nothing like a train journey to move the plot along. Holmes would know exactly where John and Joseph had been because of the qualities of the mud on their boots and that their lunch had been interrupted by urgent business because of a few crumbs in their whiskers and a telegram poking out of a waistcoat pocket.

There is another reason for making the comparison. The great detective and the former army doctor are among the world's most famous empiricists.

Figure 2.7 (a) *The Travelling Companions*, Augustus Egg, 1862.
Source: Birmingham Museums Trust.

Figure 2.7 (b) *The Third Class Carriage*, Honoré Daumier, 1863.
Source: Metropolitan Museum of Art, H. O. Havemeyer Collection, Bequest of Mrs H. O. Havemeyer 1929.

They arrive at explanations from the links in a chain of inference based on the evidence available for scrutiny. They join the dots together to solve the case, or at least Holmes does. They are not interested in speculation or the leaps of an overactive imagination. They are judged on their results. Putting their respective successes down to the power of deduction is a misnomer. To deduce, you need a theory to test. For Holmes a working hypothesis along the lines 'that this body was murdered by someone' would be facile in the extreme, worthy of Watson's misplaced common sense. 'It is a capital mistake to theorize before you have all the evidence,' Holmes reproves the good Doctor, 'It biases the judgement.'[90]

Prestwich and Evans did not like theories. Joseph, like Holmes, preferred the inductive method. Solving a problem such as the antiquity of humans required letting the facts speak for themselves by putting them in the right order. And for Joseph the 'right order' was based on the stratigraphic principles of geology. By contrast, his friend Hugh Falconer, blessed with a larger imagination, tended to jump between deductive and inductive approaches. Prestwich's younger friend, John Evans, was at this stage more impulsive, prone to deductive leaps. But that changed with age.

Joseph could have used the trip to the Somme as the opportunity to falsify, scientifically, the commonly held proposition, 'That', as he put it, 'man did not exist until after the latest of our geological changes and until the dying out of the great extinct animals'.[91] A deductive approach would set this up as a null hypothesis to be knocked over by evidence to the contrary. But Prestwich preferred, like Holmes, to build a case from the evidence rather than test a preconceived notion with evidence. Before he left for France, he was very sceptical that any reliable evidence to overturn the received wisdom would be forthcoming.[92] This was what attracted Falconer, who wrote of Prestwich that he was 'a quiet observer, of match-less sagacity and indomitable perseverance', and blest above all with a 'clenching authority'.[93] Joseph's aim, like the famous fictional detective's, was to be unfettered by theory, free as far as possible from prejudging the issue, a beacon for the rational approach that drove his geological passion. Everything that Boucher de Perthes was not.

Unfettered empiricist he may have aspired to be, but he did have a plan. Before departure he set himself four criteria to check out Boucher de Perthes's claims:

1. That the flint-implements are the work of man; 2. That they were found in undisturbed ground; 3. That they are associated with the remains of

extinct Mammalia; 4. That the period was a late geological one, and anterior to the surface assuming its present outline, so far as some of its minor features are concerned.[94]

He wanted to see and sift the evidence for himself. Holmes was always impatient with Inspector Lestrade, dubious about his powers of observation. Likewise, Joseph only believed his own eyes, which was why he travelled so much.

He was in Amiens the day before, 26 April. There he met with a local architect and antiquary, Charles Pinsard (1819–1911). Together they visited Fréville's pit at Saint-Acheul but with the same lack of success he and Evans had encountered with Boucher de Perthes on the morning of 27 April. He left instructions with Pinsard on how to contact him and to let him know if the *terrassiers* at Saint-Acheul had found what he wanted. They had, and Pinsard's telegram was the result. Perhaps in Amiens he would finally hunt down his quarry. That for him was the real contest.

John, most likely, was catching up on some sleep. Crossing the Channel the previous night, he had experienced 'as rough a passage as the strongest stomach could desire'.[95] Storm-tossed he had reached Abbeville very late, met Joseph, and gone straight to bed, only to be up before seven the next morning. Geologizing with Joseph was pure therapy; 'a couple of days' run will do me good, for I have a great deal to worry over one way and another'.[96] These worries included water, work, the paper duty, his fiancée, and children.

Elsewhere, tragedy was about to strike. The storm which so buffeted John was regathering its force in the Irish Sea. On the night of 27 April it drove the clipper *Pomona* ashore on the Blackwater Bank just off Wexford. The ship had sailed from Liverpool at 5 a.m. on 27 April, and the *Greenock Advertiser* described the tragedy of the Caldwell family, who came from the town. Their father William, a mill mechanic, was already in the United States and had paid for the passage of his wife and four children to follow him on the *Pomona*. All five died. The only survivors were the nineteen crew and three passengers who managed to launch one of the clipper's boats. The remaining crew and passengers, 394 people altogether, drowned as the ship broke up.[97]

The train pulled into Ailly, the last station before Amiens. Ten minutes later they arrived at their destination and the game was on.

In the Shadow of the Seminary at Saint-Acheul

Charles Pinsard was there to meet them. They had swapped the charming seaside style of Abbeville's railway station for a grand imperial edifice worthy of the city with France's largest Gothic cathedral.[98] Édouard Baldus photographed Amiens train shed in 1855 for an album presented to Queen Victoria as a souvenir when she visited Paris that same year.[99] The image shows five great arches, the central three capped by pediments, dark caverns into which the tracks disappear.[100] Amiens sits spider-like at the junction of the lines to Boulogne and Lille, funnelling the Chemin de Fer du Nord down south to its terminus at the Gare du Nord in Paris.

Pinsard was 40 years old and an established architect in this, his home town. His first job was in the Department of Roads and Bridges, and in 1845 he worked on constructing the line from Amiens to Abbeville which the Englishmen had just travelled. As the railway cuttings sliced through the valley, he saw first-hand the Drift geology above the chalk. He mapped the contours to plot the route, and he found a wealth of antiquities, especially from the Iron Age and Roman periods. Charles did not stay long with bridges and roads, moving to an architectural practice three years later and then branching out on his own. Outside the city walls he specialized in renovating and constructing chateaux. In the town he contributed to several public buildings, including the impressive Hospice des Incurables. The year before, 1858, saw the third edition of his exquisite map of the city on which this afternoon's destination, the seminary at Saint-Acheul, is shown set in fields a mile to the east of the station (Figure 2.8).

A keen antiquary, Charles had been following Boucher de Perthes's work and in particular the way it was pursued in Amiens by a medical doctor Marcel-Jérôme Rigollot. Born in 1786, he was a contemporary of Jacques. For a long time Rigollot was hostile to Boucher de Perthes's claims, and his friends had to drag him to see the collections in the Hôtel de Chépy so that he could write a report on them for the Academy of Amiens.[101] His conversion did not, however, take place on the road to Abbeville but rather on the way to Cagny, south-east of Amiens, where, in 1853, he went to check reports of stone tools coming from pits at Saint-Acheul. At the same time, another locality at St. Roch, just to the west of the city, was also producing stone tools. Finding disputed objects in his own backyard forced a change of mind. Rigollot's excavations found several hundred implements, and he had the sense to draft in two well-respected geologists from Paris, Charles-Joseph Buteux and Edmond Hébert (1812–90) to confirm that the

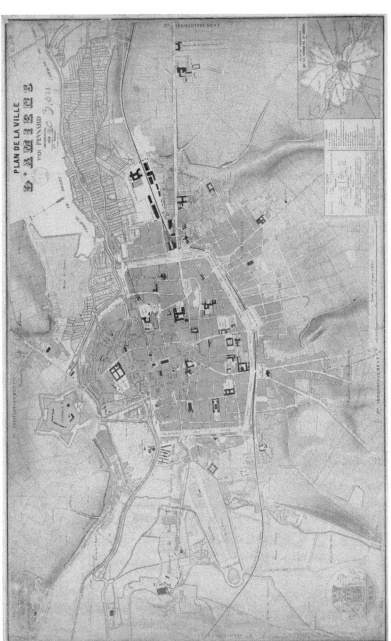

Figure 2.8 Charles Pinsard's map of Amiens in 1858. St Acheul is shown clearly on the old Roman road leading south-east out of the city.

Source: Bibliothèque Nationale de France.

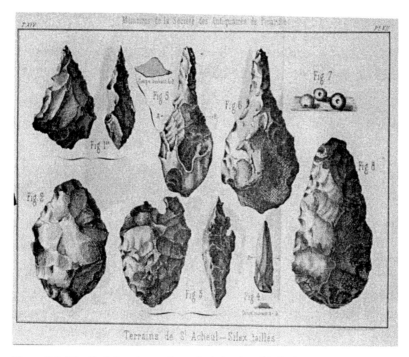

Figure 2.9 Rigollot's hand-axes from St Acheul, published in 1856 two years after his death and to a much higher standard than those of Boucher de Perthes.
Source: Rigollot 1856.

deposits they came from were indeed the Drift. His other innovation was to employ a draughtsman. The depictions of unpolished handheld axes raised the standard enormously (Figure 2.9). Compared to Jacques's tiny freehand sketches crammed onto a page (Figure 2.5), Rigollot's axes are big, half-natural-size.[102] Clearly visible on their surfaces and profiles are the scars which result from stone knapping. A firm hand had drawn these and in doing so followed the convention of depicting fossils and other natural history objects. As a result Marcel-Jérôme's axes can be discussed in terms of intention and design, whereas Jacques's scribbled sketches cannot.

In a very short time Rigollot became a convert to Boucher de Perthes's claims, writing him a mea culpa on 28 November 1854 and reading a paper on his discoveries at Saint-Acheul to the *Société des Antiquaires de Picardie* on 12 December. Then, two weeks later, on 29 December, he died.[103]

Known as 'Honest Rigollot',[104] Marcel-Jérôme would have appealed to John because of his antiquarian interests. Joseph would have recognized also a kindred spirit: good powers of observation and a reluctance to do anything but describe evidence which showed that 'men existed here at the same time as the great animals whose species were destroyed by a cataclysm'.[105]

Rigollot's paper, read to the Picardy *Société* in 1854 and published in 1856, should rightly be recognized as the proof of Jacques's antediluvian thesis. But again it was ignored. Partly because of his sudden death aged 68 so soon afterwards but also because of half-hearted support from Hébert, although Buteux gave a ringing endorsement. The crux of the matter was stated by Joseph a few years later: 'geologists admitted the antiquity of the beds, and antiquarians admitted the workmanship of the implements; but neither would own to a conjoint interest and belief in them'.[106] It was a case of putting two and two together to make four.

Pinsard was central to making that calculation. He acted as Prestwich's gofer, taking him on the morning of 26 April to the working pits at Saint-Acheul. He gave him the contour heights for his plans and sections across the valley which showed the spread of the Drift. On 27 April he assembled scientific witnesses and got them to Saint-Acheul. Above all, he arranged, at Prestwich's request, for two photographs to be taken of the flint implement, the cat's tongue, which a workman came to his house at 14 rue St Dominique to tell him about. This took place at 3 o'clock the previous afternoon and after Joseph had left for Abbeville.[107] Here was the solid proof which will make four and knock some common sense into the geologists and antiquarians.

The Englishmen and Charles arrived at the quarry pits near to a large Catholic seminary. The fact that deep history will be demonstrated in its geological basement has an antediluvian irony to it. The *terrassiers* have stopped work, and for this they will be compensated by Joseph. Laid out for inspection are the axes they have found over the last few days, and these will be purchased. Evans was overjoyed to find that in the upper layers at Saint-Acheul was a Gallo-Roman cemetery, something that Rigollot's sections showed very clearly. Stone coffins had been found, and John bought from one of the workmen a second-brass coin of Magentius (Figure 2.10). This had the letters AMB in the exergue,[108] which indicates it was struck at AMBIANUM, the Roman name for Amiens. He took the opportunity to show off his numismatic knowledge, telling everyone that the city was known to the Gauls as SAMAROBRIVA and that the earliest coins found in the cemetery date to the Claudian period.[109]

Figure 2.10 The coin Evans said that he bought from the *terrassiers* on 27 April 1859 at St Acheul, the afternoon they also discovered an *in situ* flint implement. It is characteristic of Evans that there is a slight confusion over the date when he acquired it, since the catalogue record in his own handwriting says 1860.

Source: © Ashmolean Museum, University of Oxford.

Joseph was tolerant as his best friend jumped effortlessly from the antediluvian to the historic, from the depths to the shallows of historical time. He had his tape measure out and with Pinsard's help he was measuring in the axe which the *terrassiers* had found.

Joseph jotted down the depths of the gravels and the sands of the Drift. The day before he logged sections from Saint-Acheul in his notebook, and now he added this one:[110]

Saint-Acheul
Depth of axes from surface
in 1st pit 19 ft & 21 ft

Picture—axe in—11.2"
from surface.
4.6" from bottom
of pit—
2 feet to chalk
some large subangular blocks of
sandstone are found in the gravel

The implement lay 11 feet from the surface in coarse, unsorted gravel and was 4 feet 6 inches above the bottom of the pit. A further 2 feet reached the chalk beneath.[111] In the same notebook Joseph had annotated sketches of geological sections from other pits at Saint-Acheul but not this one that he will soon have photographed. Time that afternoon was short.

The scientific witnesses arrived. In Amiens they had exchanged the *Société Imperiale d'Émulation* of Abbeville for the *Société des Antiquaires* of Picardy, today represented by its president, Charles Dufour, and secretary, Jacques Garnier, who had a large collection of elephant bones from the pits.[112] Together with John, Joseph, and Charles these were the five witnesses to what was about to happen, not to mention several workmen and no doubt curious children drawn by the buzz of activity. The five were distinguished by their tall top hats, the workmen by caps. Nobody mentions if any seminarians took an interest.

But first there was a photograph to take. The local photographer arrived with his giant camera mounted on a tripod. It had a fixed lens and a black cloth hood. A workman was instructed to stand with his finger pointing to the axe buried in the gravels, while another, grateful for the rest from digging, sat in a wheelbarrow. On the skyline the roof of a shed can be seen. Everyone was told to stand quite still as the cap was removed from the lens. The standing *terrassier* moved ever so slightly as the exposure lasted for several seconds (Figure 2.11).

The photographer whisked the glass plate out of the back of the camera. The advantages of a dry over a wet plate allowed the glass 'negative' to be taken back to the studio in Amiens and several prints made from it for Prestwich and Evans to take back to London.[113] Photography, once referred to by Henry Fox Talbot (1800–77) as the 'Pencil of Nature' in 1844,[114] but by others as the 'dark arts' (1841) and 'modern necromancy (1845)',[115] was now an instrument of history and science.

The general view taken, the camera was repositioned for a close-up of the implement in its bed of gravels (Figure 2.12), this time with no human scale, and the process repeated. And there it is poking out of its surrounding gravels just like a tongue.

These photographs are remarkable for many reasons. They are the earliest examples of a recorded event in practical, field archaeology, an event that proved humans lived in deep geological time. Quite simply, the images mark the moment when deep human history began,[116] while at the same time providing the evidence which led to its widespread acceptance. Folded into the two images are surface and deep history, a history of nineteenth-century

Figure 2.11 The time revolution captured in a photograph of the gravels at Fréville's pit on the afternoon of 27 April 1859. The *terrassier* points to the *in situ* 'cat's tongue' which can be seen in close-up in Figure 2.12.

Source: © Bibliothèques d'Amiens Métropole.

invention and modernity, simultaneously an art and a science, and a stone tool in a riverbed that was a creation of human ancestors and very old—an invention that freezes the moment, capturing the lapse of time since the flow of a river and its contents was halted. The pictures fold seconds into

Figure 2.12 The close-up of the *in situ* stone implement photographed embedded in 'higgledy-piggledy' gravels (Murchison 1861 [1808]) at Fréville's pit, just before Prestwich dug it out.

Source: © Bibliothèques d'Amiens Métropole.

millennia and vice versa.[117] Furthermore, the photographs bear witness for those not present. As John put it, 'to corroborate our testimony'.[118] The two pictures seem to state the facts without interpreting them. The capacity of the photograph to appear 'theory-free' appealed to Joseph and John. Such images were an inductive proof that then allowed deduction to take wing. All those who subsequently saw the photographs became observers of an event in a foreign field on a Wednesday afternoon in late April. Light had been captured on paper and distance had been shrunk. A moment in time could now be transported to a different time and place.

This embedded axe was the *in situ* evidence they had come to France to find. Joseph provided a detailed description:

> It was lying flat in the gravel...one side slightly projected. The gravel around was undisturbed, and presented its usual perpendicular face. I carefully examined the specimen, and saw no reason to doubt that it was in its natural position, for the gravel is generally so loose that a blow with a pick disturbs and brings it down for some way around.[119]

Like Holmes at the scene of the crime he inspected it closely and then prised it out of the gravel matrix with his pocketknife.

Once it was in his hand, he saw that 'it was the thinnest side which projected, the other side being less finished and much thicker,—a position therefore the reverse of that which would have been adopted had it been pushed in'.[120] Fraud was ruled out, and point two of his research plan had been proved. With extinct animal bones from the same levels, points 3 and 4 also followed.

Only point 1 remained: 'That the flint-implements are the work of man'. Prestwich supplied a first impression; 'This implement is rougher and more imperfect than the generality of the specimens; still it exhibits evident traces of working, especially on one side and at the point. It is an unfinished implement.'[121] In his manuscript sent to the Royal Society he also wrote 'it is not at all a specimen that would have been selected for its appearance'.[122] But this comment did not make it into print. What both he and John had hoped for was more perfect specimens such as those the *terrassiers* had found on the same day at depths of 20 and 24 feet.[123] In fact, they were surprised by the quantities that had been turned up, and when they asked, they were told that, when the pits were in full work, scarcely a day passed without these objects being found: 'The number was in fact one of the surprising features of the case' (That is Joseph, not Sherlock speaking).[124]

Although they required a long and careful search, these were not rare finds: one implement per cubic yard of gravel was Joseph's calculation.[125] Rigollot collected 400 between the months of August and December in 1854.[126] John and Joseph returned from Saint-Acheul with twenty-six axes, and every subsequent visit produced similar numbers.[127]

Point 1 of the Prestwich plan still had to be decisively answered, and that would be John's task back in England. But their day was almost done. They packed up, thanked their hosts and especially Charles, collected copies of the photographs, boarded the train, and celebrated their success with the best dinner the Tête de Bœuf could provide. The next afternoon they left for England, laden down with their stone bounty and reached the Euston Hotel by midnight.[128] Their work had only just begun.

Notes

1. Bradshaw's *Continental Railway Guide* (1853: 141) also recommended the *L'Angleterre* for travellers.
2. Prestwich 1860: 282. In a letter to R. I. Murchison, 2 April 1861, Charles Lyell encourages him to visit the Somme and singles out Marcotte and Pinsard as the people to contact (Geological Society Archives LDGSL/838/L/17/38).
3. Gowlett 2009.
4. G. A. Prestwich 1895: 944, reprinted in Milne 1901: 73–92.
5. 'Maniement d'hommes', Exposition Musée Boucher de Perthes, Abbeville.
6. For full accounts of Boucher de Perthes's life, writings, and achievements, see Aufrère 1940, Cohen and Hublin 1989, and succinct summaries by Gowlett 2009, and Sackett 2014. Dubois (2011) contrasts Boucher de Perthes with another important collector Édouard Lartet (Chs 5 and 6).
7. Sackett 2014.
8. Aufrère 1940: 19; Hublin 1997.
9. Sackett 2014: 2.
10. Aufrère 1940: 20.
11. Aufrère 1940: 14.
12. An overview of his literary and scientific output is provided by Aufrère (1940) and Cohen and Hublin (1989: ch. 2).
13. Aufrère 1940: 33–4.
14. Gill 2009: 250–3.
15. Gill 2009: 251.
16. Sackett 2014: 2.
17. Murchison 1808 [1861] LDGSL/841/25 Geological Society Archives.

18. Prestwich 1860: 2790.
19. Hughes 2017.
20. Boucher de Perthes 1860, 1842.
21. Evans 1949: 117; Boylan 1979: 178; Sackett 2014: 3.
22. G. A. Prestwich 1895: 944.
23. These three qualities of time are described by Corfield (2007) as shaping the course of history.
24. Prestwich 1860: 282.
25. Thackray 2004: 276.
26. French was the language of international relations. Up until 1858, if anyone wanted a British passport, and few bothered, it was written in French not English.
27. Brickearths are windblown loessic deposits reworked by river action to form substantial beds of clay. Loess formed in the cold climates of the ice ages when sparse vegetation in front of the ice sheets led to massive wind erosion followed by deposition. Brickearths are one source of the clays to make bricks.
28. Evans 1860a: 283.
29. See Evans 1949 for a full description.
30. Picard 1834-5, 1836-7; Ravin1834-5.
31. M. F. Aufrère 2012. See also the Catalogue Collectif de France for details of those involved: http://ccfr.bnf.fr/portailccfr/jsp/public/search/public_search_result_ exec.jsp?page=4&source=bmr&pagerName=search, accessed 27 March 2018.
32. Grayson 1983: 120.
33. Boucher de Perthes 1847: 146.
34. Grayson 1983: 126.
35. Boucher de Perthes 1847: 236 and 348.
36. Boucher de Perthes 1847 Plate XVII. Joan Evans reflects the concern that many had that some of the much younger Neolithic, axes were planted by the *terrassiers* in the older deposits (Evans 1949).
37. Aufrère 1940: 69, Figs 7 & 8.
38. Boucher de Perthes 1847: 348.
39. Champ de Foire (Boucher de Perthes 1847).
40. Boucher de Perthes 1847: 247.
41. Grayson (1983: 122) has the full story of why Jacques delayed publication by two years.
42. Prestwich 1860: 283.
43. Boucher de Perthes 1847: 253.
44. Prestwich had used this method previously to trace a fossil beach at 125 feet above sea level as it appeared in pits across the chalk downs of Sussex and Hampshire (Prestwich 1859b).
45. Buteux 1857. This was published in the *Mémoires de la Société d'Émulation d'Abbeville*, although due to the collapse of the monarchy, it was now an Imperial rather than Royal Learned Society.

46. Prestwich 1860: 283 n 1.
47. Prestwich 1860: 283–7.
48. Smith 1888: 7.
49. Boucher de Perthes 1857: 108.
50. Perry 2011: 178. Smith and Eaton were particularly active forgers between 1857 and 1860, when they salted the Thames foreshore.
51. Evans 1865; Lamdin-Whymark 2009. 'Flint Jack' had as many pseudonyms as a highwayman. They included 'Fossil Willy', 'Shirtless', and 'Old Antiquarian', as recorded in the *Whitby Gazette* in 1867, when he was sentenced to a year in jail. See https://en.wikipedia.org/wiki/Edward_Simpson_(forger), accessed 25 July 2020.
52. See Glossary. Geologists declared in 2016 that we now live in the Anthropocene (see Chapter 7), which follows Lyell's Holocene.
53. Prestwich 1860: 286.
54. Prestwich 1860: 287.
55. Prestwich 1860: 286.
56. Prestwich 1860: 287.
57. Evans 1943: 101.
58. Murchison 1808 [1861] LDGSL/841/25 Geological Society Archives.
59. *Musées* 80 no. 2, p. 13, reproduced in Cohen and Hublin 1989: 231.
60. Prestwich 1859–60: 53.
61. Aufrère 1940: 12, Fig. 2.
62. G. A. Prestwich 1895: 943.
63. See Chapter 6.
64. Evans 1949: 121.
65. Falconer wrote to Boucher de Perthes announcing their impending visit in October 1858. He had first met Jacques in 1856 (Falconer 1868: 596). His account, published posthumously, was written in 1863.
66. G. A. Prestwich 1895, reprinted in Milne 1901: 73–92.
67. G. A. Prestwich 1895: 943.
68. G. A. Prestwich 1899: 119–20. Letter dated 1 November 1858. Jacques became the head of his family after the death of his father in 1844. His elder brother Jules had died two years previously.
69. G. A. Prestwich 1895: 944.
70. G. A. Prestwich 1899: 119. See Richard (2011) for the importance of these learned societies in France and their role in the 1859 time revolution.
71. Grayson 1983: 129. *Darwin Correspondence* In a letter to Hooker on 22 June 1859, after Evans and Prestwich had presented their evidence, Darwin was still reluctant to accept Boucher de Perthes's stone tools. In Darwin's opinion they were formed by ice action. He later recanted when the evidence was explained to him by Lubbock.

72. The Société still flourishes in Abbeville: see http://www.societe-emulation-abbeville.com/, accessed 25 July 2020.

73. Boucher de Perthes 1847: 261–2; Cohen and Hublin 1989: 120–1.

74. Aufrère 1940: 82 and Figs 12 and 13; Boucher de Perthes 1847: 351 There are some complications with the descriptions and this hand-axe came from a fluvial deposit somewhere between the Hôpital and the old Fairground near the rue Millevoye.

75. Prestwich 1860: 279; Evans 1860: 294. There are similar *procès verbaux* in Boucher de Perthes (1857: 430 and 459).

76. See Evans (2012) for a discussion of archaeology as an experimental rather than a discovery science.

77. Prestwich 1859–60: 56.

78. Boucher de Perthes 1847: Plate LXIII and p. 485.

79. Prestwich 1859–60: 53.

80. Newton was also a passionate mythographer who revised classical and biblical chronologies (Kidd 2016: 81).

81. Wallace 1898. Phrenology is the pseudoscientific study of head shapes to reveal a person's character. It was hugely popular in the nineteenth century.

82. See Piggott (1968) for a forensic deconstruction of Stukeley's creation.

83. http://www.aod-uk.org.uk/home.htm, accessed 27 September 2016.

84. See Chapter 7.

85. Murchison 1808 [1861] LDGSL/841/25 Geological Society Archives.

86. The telegraph is one of the nineteenth century's many wonders. The story of its invention is complex. The discovery that electrical impulses could be transmitted over distances, sent as code, and translated back into words started forty years before. The pace accelerated in the 1830s, with inventors in Germany, England, and the United States all coming up with different solutions. Two systems were both patented in 1837: one by the Englishmen William Cooke and Charles Wheatstone, and the other by the American Samuel Morse with his eponymous code key. The end result was the same: time had collapsed distance, making it smaller. From 1844 messages started to flow along the wires with Morse's sententious 'What Hath God Wrought!' (perhaps he should have ended with a question mark?) sent from Washington to Baltimore. Such communication was commonplace by 1859 and by coincidence 27 April was Morse's sixty-eighth birthday.

87. Dickens and Collins 1857: Kindle p. 41.

88. Bradshaw 1853. If they had taken the express, they would have covered the 30 miles in 45 minutes.

89. https://www.youtube.com/watch?v=7XPEoiY3NnI, accessed 25 July 2020.

90. Conan-Doyle *The Study in Scarlet* (1887 [1929]: 28).

91. Prestwich 1860: 2, where he presents it as the established wisdom.

92. Prestwich 1859–60: 51.

93. G. A. Prestwich 1899: 123.
94. Prestwich 1859–60: 58; 1864a: 247.
95. Evans 1943: 101.
96. Evans 1943: 101.
97. http://thegreenockian.blogspot.co.uk/2014/06/sinking-of-pomona-1859.html, accessed 19 September 2016.
98. Abbeville's railway station is a fine example of the seaside style and was opened in 1861.
99. The Queen's Album is kept in the Royal Library at Windsor Castle. It was commissioned by Baron James de Rothschild, who was President of the Chemin de Fer du Nord.
100. Braun and Kingsley 2015: 87.
101. Grayson 1983: 173; Boucher de Perthes 1857: 467–8.
102. Rigollot 1856: Plate VII.
103. Grayson 1983: 175; Prestwich 1864b: 215.
104. http://www.inha.fr/fr/ressources/publications/publications-numeriques/dictionnaire-critique-des-historiens-de-l-art/rigollot-marcel-jerome.html, accessed 25 July 2020.
105. Grayson 1983: 175; Rigollot 1854: 3–4 and 1856: 23–4.
106. Grayson 1983: 175; Prestwich 1864b: 215.
107. Pinsard's account contained in his papers in the Amiens Bibliothèque is difficult to follow in terms of dates and times.
108. A small space or inscription below the principal emblem on a coin or medal, usually on the reverse side.
109. Evans 1860a: 285.
110. Prestwich 1850–9: 115.
111. Prestwich 1860: 290–1; Evans 1860a: 295.
112. Prestwich 1860: 291 n. Hurel and Coye 2011. Lyell, *Letter* to Murchison, 2 April 1861. LDGSL/838/L/17/38 Geological Society.
113. When Georges Pouchet visited Saint-Acheul on 28 August, he was given copies of the photographs (Pouchet 1860, in de Bussac 1999: 300). Photography was evolving very rapidly, moving from a professional to popular pursuit as the process was simplified. In 1862 Professor Pole, fellow of the Royal Society, had some handy tips for travelling photographers. Instead of carrying around 'a fearful array of dark rooms, huge instruments, chemical paraphernalia, water and mess' the dry plate reduced the load enormously. According to Pole, all that the traveller needed was a five-inch-square case of bottles, and a stock of glass plates, a dozen of which fitted into a box $8 \times 4 \times 1^{1/2}$ inches. The camera had also reduced in size to a more manageable 2 lb in weight. But it still needed a fold-up stand. Pole's claim that the traveller could now walk about without the slightest sense of encumbrance seems optimistic (Pole 1862: 250).

114. Fox-Talbot 1844–6. Michael Faraday made the announcement to the world that photography existed in a lecture to the Royal Institution, London on 25 January 1839. Two pioneers, William Henry Fox Talbot in England and Louis Daguerre in France, devised related but different techniques to make photographs.
115. Roberts 2016: 19–20, citing Herschel in 1841 and the *Athenaeum* in 1845.
116. Cohen and Hublin 1989: 185.
117. LaTour 2005: 201.
118. Evans 1943: 102.
119. Prestwich 1860: 291–2.
120. Prestwich 1860: 292.
121. Prestwich 1860: 292 n.
122. Prestwich 1859a, : 42.
123. Prestwich 1860: 292.
124. Prestwich 1860: 292–3.
125. Prestwich 1864b: 216.
126. Prestwich 1860: 293.
127. Prestwich 1860: 293 n.
128. Evans 1943: 102.

3

Presenting the Evidence

The Month May – June 1859

Two days after his return from France Joseph Prestwich sat down at his desk in Kent Terrace and dashed off letters to a select group of friends and geologists. They were invited to inspect the evidence brought back from the Somme and drink some of his 'legendary sherry'.[1] The gathering, a Flint soirée, took place at 8.30 p.m. on Tuesday 3 May. Civil Mary, his sister, played hostess to an all-male event. Evans took his friend Augustus Wollaston Franks (1826–97), Director of the Society of Antiquaries and a curator at the British Museum, as was George Waterhouse, who also attended.[2] Three other likely invitees would have been the founding members of the London Clay Club, which in 1846 transmuted into the Palaeontographical Society.[3] The three appear alongside Joseph in an undated photograph of a 'flint conference' published by Grace Prestwich in the biography of her husband in 1899 (Figure 3.1).[4] Joseph holds an antediluvian axe. Looking over his left shoulder is John Morris (1810–86), Professor of Geology at University College London, and at the far right is Searles V. Wood (1798–1880), who, like any good palaeontographer, wrote weighty monographs on the Crag Mollusca from Neogene deposits in Suffolk and Norfolk.[5] The last Clay Clubber, Frederick Edwards (1799–1875), is awkwardly holding two more flint axes.[6] Edwards's obituary mentioned a common affliction for Victorian men of science. He was 'often harassed by family cares and anxieties, and oppressed with the responsibilities of his official work as a solicitor'.[7] Stresses that Joseph and John also felt keenly but were able, for the most part, to mitigate successfully.

All three were old friends of Joseph's. They dealt in descriptions and facts, not theory. The photograph may not commemorate the first view that English scientists had of artefacts from the Pleistocene Drift of the Somme. Nonetheless, it captures the male camaraderie of Victorian science and the interest generated by what John and Joseph had found but which lay outside their palaeontographical passions. The good opinion of the Clay Club mattered.

Figure 3.1 *A conference on Flint Implements.* The picture is undated and all we have to go on is how old the participants appear to be. In 1859 Prestwich, far left, was the youngest of the group at 47, and Searles Wood, far right, the oldest at 61. A date in the early 1860s is most likely.

Source: G. A. Prestwich 1899.

The day after the soirée Evans left on a week-long business trip. First stop was Shrewsbury via Wroxeter, where he visited the Roman town that the antiquary Thomas Wright (1810–77) was digging incompetently.[8] On 5 May he was in Dublin seeing clients and then on to Drogheda and Belfast, taking the opportunity on the Sunday to visit more antiquities in Antrim, Ballymena, and Loch Neagh. He travelled on Tuesday from Belfast to Manchester, a city important both as a market for Dickinson's and as a source of rags, the raw material for making paper. He arrived back in Nash Mills the next day, Wednesday, 11 May. On Thursday he was at Somerset House in London at the Society of Antiquaries.[9]

Two Learned Societies

While John was clocking up his train miles, Joseph started on his paper. Two audiences needed to be won over: scientists and antiquaries. The obvious first choice was the Royal Society, whose fellows had a wide breadth of scientific interests. Convince them and lower-ranked bodies such as the Geological Society,[10] to which they both belonged, would follow.

In 1859 the Royal Society was one-year shy of its two-hundredth anniversary.[11] Joseph had been a fellow since 1853, and among those who signed his ballot paper, citing personal knowledge of his scientific abilities, were the eminent geologists Sir Charles Lyell and Sir Roderick Murchison, and the naturalist Thomas Huxley. The motto of the society must have appealed to Joseph, 'Nullius in verba', which translates as 'Take nobody's word for it.' Statements have to be verified by facts determined by experiments that can be repeated. At the same time, the motto exhorts fellows to speak the truth and resist pressure to do otherwise from politicians, churchmen, and any other interest group outside science.

Their second target, the Society of Antiquaries, was determined by the archaeological interest of the flint implements. This learned society was founded by three friends who met at the Bear tavern in the Strand on 5 December 1707. By their second meeting a week later they had agreed the business of the society to be limited, among other things, 'to the subject of antiquities; and more particularly, to such things as may illustrate and relate to the history of Great Britain'.[12] Their chronological sweep took in everything before James I came to the throne a century before. Wisely they set a cut-off for their meetings of 'Ten of the Clock at the furthest' and agreed that fellows only had to pay for what they ordered at the bar.[13]

The Antiquaries grew rapidly, gaining its Royal Charter in 1751 and moving into Somerset House alongside the Royal Society in 1781. Their names can still be seen carved over their respective doors, although both societies are now housed elsewhere.[14] From this palatial base antiquaries set out to encourage the 'ingenious and curious'.[15] Their meeting room was arranged like the debating chamber of the House of Commons, with a central large table and opposing benches. It was around this table that the past was seen, handled, and talked into existence.[16] Over the years the objects on the baize-covered table took many forms: classical vases, brass rubbings and other church fittings as well as heraldic paraphernalia, coins, medals, weapons, and manuscripts. In his cartoon of 1812 George Cruickshank satirized the Society by adding a coal scuttle, labelled as a recent shield, a chamber pot with the ticket 'Roman vase' attached, and numerous scabrous jokes at the fellows' expense.[17] Those who studied the past were known as antiquaries, the designation 'archaeologist' rarely used. Prestwich and Evans needed their endorsement of the stone implements from Saint-Acheul.

In 1859 antiquarian studies were a gentlemanly pursuit rather than a systematic endeavour. Archaeological excavations rarely employed geological observations as detailed as those provided by Prestwich for the Somme. Digs were just that, diggings like Wright's at Wroxeter, a search for curiosities, with scant attention paid to their stratigraphic context.

Two Short Abstracts, a Letter, and Two Long Papers

The scientific responsibilities of the two friends were clear. Prestwich, as senior author in the senior learned society, dealt with the geological circumstances of the finds at Abbeville and Amiens. While John was in Ireland, Joseph started to sketch out his paper to be read to the Royal Society later that month. He was buoyed by the positive reaction of those who came to Kent Terrace to see the artefacts and geological sections. It will be down to Evans to tackle the tricky matter of presenting these tools as irrefutable evidence of human handiwork. By 1859 John had been elected to three learned societies: the Numismatic Society since 1849, the Antiquaries in 1852, and the Geological Society, where Prestwich had proposed him in 1857.[18]

Publication in 1859 was a slow business, not because of the mechanics of printing but rather because papers had to wait for others, including editors, before they could go into the monumental journals of record. These weighty tomes were the *Philosophical Transactions* for the Royal Society and

Archaeologia for the Antiquaries. The solution was to publish short abstracts in the *Proceedings* of both societies. This got the word out faster and established claims to knowledge and discoveries.[19]

Joseph's abstract appeared in volume 10 of the *Proceedings of the Royal Society of London*. This left time for the longer paper to be revised, refereed, and published the next year in the *Transactions*. His title, submitted to the Royal Society on 19 May, made no reference to where the finds were made; *On the Occurrence of Flint-Implements, Associated with the Remains of Extinct Mammalia, in Undisturbed Beds of a Late Geological Period*. John's title follows almost word for word, *On the Occurrence of Flint Implements in Undisturbed Beds of Gravel, Sand and Clay*. His abstract, read on 2 June, was published in volume 4 of the *Proceedings of the Society of Antiquaries*. Amiens, Abbeville, and Hoxne were tacked on by Joseph to the title of his longer paper in the *Transactions*. John did not follow suit in *Archaeologia*.

Two short abstracts and two lengthy papers of record announced the arrival of deep human history. How did they make their case? Both authors dealt with the background to their inquiry and then presented the geology and flints in a well-rehearsed, agreed account of the events on the Somme.[20] They borrowed freely from each other, so that it is like reading a joint paper, cut and pasted, where individual authorship dissolves. Only occasionally do they re-emerge. A ten-page letter by Evans about the flints, written from Nash House on Dickinson's best blue paper, was sent the day before Prestwich gave his talk to the Royal Society. It was included with Joseph's *Transactions* paper as Appendix A. Literally cut and pasted. It survives in the Royal Society's library.

Their tone throughout is self-deprecating, appealing to rational opinion as reasonable, businesslike men. Joseph confessed 'that he undertook the enquiry full of doubt'.[21] First they had to convince themselves. John wrote in 1868, 'I who did not believe in the slightest degree in the probability of this reputed discovery, was obliged to accept it as a genuine and most important one.'[22] Then they had to convince others. Prior to their discovery in 1859 Joseph believed that scarcely a geologist in the country believed it possible for humans to have a great geological age.[23]

Both of them expounded on the implements as human handiwork and both acknowledged their debt to the work of Boucher de Perthes, Rigollot, Pinsard, Ravin, and Buteux. Neither man attempted to drape the British flag of science over the Frenchmen's work, presenting their results instead as an international endeavour.

Two Parallel Cases: Deferred and Unexpected

With their new-found evidence what they now wished for was a parallel case to corroborate their argument. And here there was a stumbling block, Brixham Cave in Devon. Quarrying for stone had brought this bone cave to light in 1858. Such discoveries were usually soon accompanied by visits from palaeontologists such as William Pengelly and Hugh Falconer in search of ice age bones of mammoth, cave bear, reindeer, and woolly rhino. Brixham was first and foremost a geological cornucopia.[24] But it also produced stone tools comparable to finds from caves reported from Europe, at Liège by Dr Phillipe-Charles Schmerling (1790–1836) and in France by Marcel de Serres (1780–1862). There were also finds from Kent's Cavern, a large cave in Torquay, where pioneering excavations by the Reverend John MacEnery (1797–1841) during the 1820s had yielded a clear association between stone tools and extinct animals. However, MacEnery came under the sort of pressure that elsewhere fellows of the Royal Society were exhorted to resist, and in time came to renounce the association.[25] Prestwich's *Transactions* paper generously paid tribute to his work.[26] He had this luxury because he was writing after MacEnery's nemesis, William Buckland of Oxford, had died. Buckland used geology to support the chronology of the Bible and in particular the catastrophe of Noah's flood.[27] His death in 1856 had removed one of the biggest obstacles to their evidence from the Somme being accepted. Joseph used this to his advantage, declaring in MacEnery's favour, 'I do not see the grounds on which the evidence was objected to, except that the fact was considered impossible and the chances of error great.'[28] More to the point, no specific grounds were ever put forward, just opinions and theories like Buckland's which prejudiced acceptance of MacEnery's well-made scientific case.

In his abstract for the Royal Society Joseph commented on the work of the Brixham Committee of which he was a mainstay. He also mentioned a preliminary paper that Uncle Hugh had already presented to the Royal Society, where he claimed that at Brixham there were worked stones 'mixed indiscriminately with the bones of the extinct cave bear and the Rhinoceros'.[29] But Joseph was not convinced by the Brixham finds, or for that matter any other cave evidence. The sticking point was contamination. Caves were heavily used in later periods. Pits were often dug into them. These could mix up the sediments and, worse, introduce artefacts into layers to which they did not belong. For Prestwich and the others on the committee there were simply too many doubts.

Downplaying the Brixham evidence was a clever strategy. It showed scientific integrity. Unintentional contamination, scientific fraud, and the possibility of forged implements were at a stroke swept aside, as might be expected from Joseph, a geologist with 'clenching authority'.[30] By removing Brixham from the pool of evidence, Prestwich was eliminating one controversial claim and replacing it with *in situ* photographed and accurately recorded evidence. His strategy explains why Brixham, where he and Pengelly found stone tools the year before, was abandoned and only published much later in 1873, when it was no longer controversial.

Prestwich's abstract was received by the Royal Society on 19 May, eight days after John returned from Ireland, and read at the weekly meeting of fellows on Thursday, 26 May. But between John's return and the paper arriving at the Royal Society events took an unexpected twist that made their task easier.

The Society of Antiquaries meets on Thursday afternoons, and John was waiting in their rooms at Somerset House on 12 May for the meeting to finish and a friend to come out. It is likely to have been Augustus Franks of the British Museum, who would have been attending as the society's director. John idled away the time by glancing into the familiar glass-fronted cases that contained objects which fellows had donated. These were as varied as the interests of the fellowship: curiosities, antiques, coins, medieval crucifixes, silver cups, floor tiles, paintings, and bronze shields. And then he stopped: 'I looked at a case in one of the window seats, and was absolutely horror-struck to see in it three or four implements precisely resembling those found at Abbeville and Amiens.'[31] Up until that moment he had assured Joseph that what they had found in April was unlike anything known in England. But there they were: unpolished axes worthy of Boucher de Perthes's *haches antédiluviennes* gathering dust in London.

The axes were unlabelled, and no one knew anything about them. Undaunted, John dug into the archives in the Antiquaries Library, and here the meticulous Franks would have been a great help. What came out was gold dust. The axes had been sent to the Society in 1797 by one of its fellows, John Frere (1740–1807), a Suffolk landowner. He accompanied them (Figure 3.2) with the most extraordinary account of how they were found and what he believed their significance to be. Where Jacques would never use a hundred words when there was room for a thousand, Frere's brief letter spoke volumes. Figure 3.2 shows his short communication addressed, sixty-two years previously, to the Antiquaries' secretary, the Reverend John Brand.

XVIII. *Account of Flint Weapons difcovered at* Hoxne *in* Suffolk. *By* John Frere, *Efq.* F. R. S. *and* F.A. S. *In a Letter to the Rev.* John Brand, *Secretary.*

Read June 22, 1797.

Sir,

I TAKE the liberty to requeft you to lay before the Society fome flints found in the parifh of Hoxne, in the county of Suffolk, which, if not particularly objects of curiofity in themfelves, muft, I think, be confidered in that light, from the fituation in which they were found. See Pl. XIV, XV.

They are, I think, evidently weapons of war, fabricated and ufed by a people who had not the ufe of metals. They lay in great numbers at the depth of about twelve feet, in a ftratified foil, which was dug into for the purpofe of raifing clay for bricks.

The ftrata are as follows:

1. Vegetable earth 1½ feet.
2. Argill 7½ feet.
3. Sand mixed with fhells and other marine fubftances 1 foot.
4. A gravelly foil, in which the flints are found, generally at the rate of five or fix in a fquare yard, 2 feet.

In the fame ftratum are frequently found fmall fragments of wood, very perfect when firft dug up, but which foon decompofe on being expofed to the air; and in the ftratum of fand, (No. 3,) were found fome extraordinary bones, particularly a jaw-bone of enormous fize, of fome unknown animal, with the teeth remaining in it. I was very eager to obtain a fight of this; and finding it had been carried to a neighbouring gentleman, I inquired of him, but learned that he had prefented it, together with a huge thigh-bone, found in

in the fame place, to Sir Afhton Lever, and it therefore is probably now in Parkinfon's Mufeum.

The fituation in which thefe weapons were found may tempt us to refer them to a very remote period indeed; even beyond that of the prefent world; but, whatever our conjectures on that head may be, it will be difficult to account for the ftratum in which they lie being covered with another ftratum, which, on that fuppofition, may be conjectured to have been once the bottom, or at leaft the fhore, of the fea. The manner in which they lie would lead to the perfuafion that it was a place of their manufacture and not of their accidental depofit; and the numbers of them were fo great that the man who carried on the brick-work told me that, before he was aware of their being objects of curiofity, he had emptied bafkets full of them into the ruts of the adjoining road. It may be conjectured that the different ftrata were formed by inundations happening at diftant periods, and bringing down in fucceffion the different materials of which they confift: to which I can only fay, that the ground in queftion does not lie at the foot of any higher ground, but does itfelf overhang a tract of boggy earth, which extends under the fourth ftratum; fo that it fhould rather feem that torrents had wafhed away the incumbent ftrata and left the bog-earth bare, than that the bog earth was covered by them, efpecially as the ftrata appear to be difpofed horizontally, and prefent their edges to the abrupt termination of the high ground.

If you think the above worthy the notice of the Society, you will pleafe to lay it before them.

I am, Sir,

with great refpect,

Your faithful humble Servant,

JOHN FRERE.

XIX. *Ac-*

Figure 3.2 Frere's short letter to the Society of Antiquaries about the finds at Hoxne, Suffolk. Sent in 1797 and published in *Archaeologia* three years later. The two hand-axes are shown in Figure 3.5.
Source: Frere 1800.

John Frere's handwritten letter does not survive. It was published in *Archaeologia*, the Antiquaries' journal of record, in 1800, and then promptly forgotten about. Evans saw at once that here was a most important parallel case to the discoveries in France by Joseph and himself. And where had this taken place? Outside the small Suffolk village of Hoxne in a pit being dug for clay to make bricks. One phrase in the letter made his chance rediscovery even more interesting, that the quarrymen had emptied baskets of these implements to fill up the ruts in the road. These were not rare finds. There must be more to discover.

John's excitement was barely containable, and he told Joseph as soon as he could. Domestic and business commitments prevented them from going straight away to Hoxne to see for themselves. The frustration surfaced in Joseph's letter to John a week later: 'I shall be restless until I visit Hoxne,' he wrote, and then suggested they visit on the weekend of 21 and 22 May, four days before he was due to speak at the Royal Society. Joseph ended the letter, 'I have found three brick pits at or near Hoxne, and hope to find traditions of discovery and to have a trench dug on the right spot.'[32]

They made that visit together, and it clinched their argument. They spoke to an old man who had worked in the pit where Frere made his discoveries.[33] As with the Amiens *terrassiers*, he confirmed the abundance, the depth, and the association of implements with extraordinary animal bones. The old man even had a name for the flint implements, 'fighting stones', to rival his French counterparts 'cats' tongues'.[34] The name echoed Frere's original description of 'weapons of war'. So there we have it. The geologist and antiquary were able to compare traditions of gravel digging that gave us, independently, a sweet French biscuit and an English flint fist.[35] The time revolution had been confirmed in less than a month. An archaeology of deep human history was now possible.

Hoxne was the independent test they craved. Their trenching in the pit confirmed Frere's observations on the geology. Significantly, in Evans's telling phrase in his blue-paper letter to Prestwich, the observations were made by 'an antiquary unfettered by geological theories'.[36] Frere's testimony can be believed because he passed Joseph's inductive test. He was not out to prove a point and prejudge the issue. Rather, his interest lay in presenting the evidence. '*Nullius in verba*'; and yes, Frere was also an FRS as well as an FSA and, to cap it all, the Member of Parliament for Norwich. This was no small-town savant like Jacques, kept at arm's length by the metropolitan elite. Frere was one of us.

The Talks

Prestwich got to his feet on 26 May to read his abstract. Nerves had got to him. His normally booming voice was cracked, and he could barely be heard in the Royal Society's rooms at Burlington House in Piccadilly. He had brought the implements from Abbeville, Amiens, and Hoxne, the diagrams of the heights of the pits across the Somme and Waveney rivers, animal bones, shells, the volume of *Archaeologia* open to Frere's account with its exquisite engravings of the implements, and no doubt other relevant papers and books from Buteux, Rigollot, and Boucher de Perthes. The flint and its two photographs were there for fellows to see and handle. He summarized Jacques's claims for axes from the Drift dividing them into two principal shapes, pointed and oval.

The meeting was well attended, but not packed. Evans described the outcome. His friend had 'dexterously managed to leave behind him' the letter John had written the day before about the flints. To his consternation he was called on to extemporize in front of a distinguished audience, in his phrase, 'of geological nobs'—Lyell, Murchison, and Morris. Also in the audience were fellows of the society: the naturalist Huxley, and other polymathic scientists who defined their age through their inventions: Michael Faraday (1791–1867), founder of the science of electromagnetics and electrochemistry, and Charles Wheatstone (1802–75), physicist and pioneer of telegraphy. Altogether an audience of high quality if not quantity. Joseph's paper was followed by one from the keen geologist, Charles Babbage, better known as the mathematician credited with the first computer. Babbage told the audience that he was perfectly satisfied that the flint implements from Amiens, and those found by Falconer in the Sicilian caves, were human handiwork. A ringing endorsement from a critical authority.[37]

John thought his off-the-cuff speech went well, mostly because it was still fresh in his mind. He added a third category of implements, flint flakes interpreted as knives and arrow heads (Figure 3.3).[38] He compensated for the croaking Joseph, who gave only an 'indifferent abstract'. By the end John was pleased: 'our assertions as to the finding of the weapons seemed to be believed'. Then he went straight on afterwards to an Alpine soirée at the home of the family friend, the publisher William Longman, 'so altogether I had rather a busy evening'.[39]

In the audience was a young man, John Lubbock, a naturalist and banker. He was impressed by what he heard and after the meeting asked Prestwich if he could join him on a future visit to Amiens.[40]

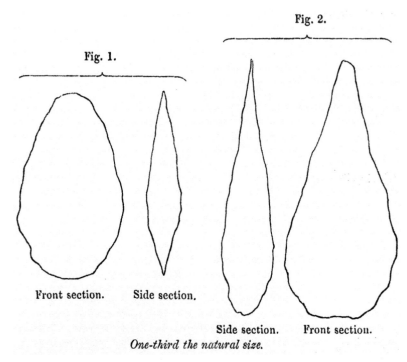

Fig. 1.

Front section.   Side section.

Fig. 2.

Side section.   Front section.
One-third the natural size.

Figure 3.3 The pointed and oval flint instruments as presented by Joseph in his abstract. These are not the first British illustrations of artefacts from deep history. Hoxne has that distinction. But they are the first time that Boucher de Perthes's claims of antediluvian axes [*haches*] were shown, approvingly, to a learned society in London. In his blue-paper letter Evans suggested that the oval forms, which he glossed as almond-shaped, were either axes or slingshots and the pointed examples were hafted as lance or spear heads. He would soon row back from these functions inspired by Boucher de Perthes.
Source: Prestwich 1859–60.

Evans does not mention Lubbock's presence at the Royal Society on 26 May, probably because they were already well acquainted, and only new encounters, preferably with celebrities such as Thackeray, were passed on to Fanny. Lubbock later recalled how he had joined the Geological Society in 1855 and come to know both Evans and Prestwich.[41] He was there in the audience doing a favour for a fellow entomologist by standing in to read his paper.[42]

Throughout the year Darwin was working hard to complete *The Origin* and called on Lubbock to answer detailed questions about insects.

On 8 February, Lubbock wrote back to him about the anomalous internal organs of *Pulex*, a genus of insects among which are numbered the human flea.[43] Darwin wanted him to review their classification using these organs because 'their externals have been discussed *ad nauseam*' and 'no one but you in England' can write such a paper.[44] Lubbock duly did this later in the year.[45] In March he was answering Darwin's queries about the metamorphoses of insects and the embryology of *Aphis*.[46] In his reply the next day Darwin thanked him for correcting his error. This constant exchange between two friends, a friendship that began as mentor and pupil, was part of the final tinkering and proof corrections for *The Origin*, about which Darwin was nervous and excited in equal measure. His internal thoughts were soon to be publicly scrutinized like the ovarium of *Pulex* under Lubbock's dissecting knife.

John's big day at the Society of Antiquaries was 2 June. Another meeting with a low attendance. The visitors' book lists thirty-eight fellows and their guests. Other lectures that year drew audiences of between twenty-three and sixty-seven, so it could have been worse, but hardly the numbers we might expect a time-changing event to attract. John was also hoarse but managed to get through it in three-quarters of an hour. What he said was 'very much admired and I think generally believed in'.[47]

Joseph was not there. He had returned to France with three other geologists.[48] One of them, John Wickham Flower (1807–73), a solicitor by profession, an antiquary and geologist by pursuit, had his fifteen minutes of geological fame.[49] In the kindly obituary Joseph wrote for his old friend he recalled how Flower with his usual zeal spent a whole day at Saint-Acheul, digging with pickaxe and spade into the gravels until he found an *in situ* flint implement at a depth of 16 feet. This was duly brought back and presented to the Geological Society on 22 June. A two-page publication followed, with a handsome, full-page engraving of the implement.[50]

The Language of Flints

How do you convince an audience of the reality of something which, although it can be imagined, still has to be proved? In the absence of human remains they had to put forward a proxy for the inhabitants of deep history. These proxies were stone tools. It is the novelty of the flint implements from the Drift that makes John and Joseph's presentation remarkable. It is like being asked to believe in an extraterrestrial from the imprint of their spacecraft left behind in the sand. Yes, John and Joseph had the French stone implements, but how would they be viewed by London's learned societies? They bolstered their case by using the practical testimony of *terrassiers* and quarrymen. They looked around the world for modern examples. John turned to the Americas and compared some objects recently found by Uncle Hugh in a Sicilian cave to obsidian knives from Mexico. And these, he told the readers of *Archaeologia*, were the tip of an iceberg. Such knives could be found throughout the world.[51]

But when it came to claiming that the implements from the deep Drift were indeed human artefacts, Evans was faced with a problem. What they had found at Amiens was embedded in the certainties of geology. But once they were removed from the gravel matrix, there was no vocabulary to describe the flint implements.

Prestwich confronted the issue with three crucial questions:

1. Were these flint implements made 'by the hand of man'?
2. Could they have been made more recently?
3. Were they contemporary with the geological beds where they were found or were they introduced from later periods?[52]

The first was by far the most important, because it stepped outside the trusted methods of geology which convincingly proved the two other points. Joseph was aware of the importance of his first question, so he set out carefully the steps in his reasoning: 'The argument', he reminded his readers, 'does not rest upon the evidence of skill, but upon the evidence of design.'[53] The manufacturing skill was 'rude' because the flints were chipped, not ground down and polished into shape. What had shaped these oval and pointed flints was 'blows applied by design, and with a given object in view'.[54] In other words, whoever made them had an intelligent plan and followed it through. He hammered this point home: 'it is ... upon the unity of character and evident object in the design, that the argument of artificial make is the strongest'.[55] Evans supported his friend's conclusion in *Archaeologia*: 'There is a uniformity of shape, a correctness of outline, and a sharpness about the cutting edges and points, which cannot be due to anything but design.'[56]

Evans could have been echoing the art historian and critic John Ruskin (1819–1900) who wrote in his influential *Stones of Venice* in 1853, 'the rule is simple: always look for invention first, and after that, for such execution as will help the invention'.[57] Consequently, Ruskin was unbothered by flaws in the execution of a craft, claiming there is no reason 'to be proud of anything that may be accomplished by patience and sand-paper'.[58] In his attack on the machine-made object, a point on which Evans the papermaker would have disagreed, he continued: 'accept then for a universal law, that neither architecture nor any other noble work of man can be good unless it be imperfect'.[59] Applied to the stones of Saint-Acheul, Ruskin's argument about roughness and design, if John had used it, would have praised the authenticity of Jacques's coarse axes and questioned the polished ones as merely the work of perseverance and elbow grease.

An appeal to Ruskin might have helped in some circumstances, but what knowledge could John call on to convince a sceptical audience of scientists and antiquaries? Two solutions have been put forward. What Evans did in the face of uncertainty was to fall back to his comfort zone, his business experience of industrial patents and his studies of ancient coins, and fold them into the deep history he was creating.[60] He did not do this consciously, but it was his way of setting out to prove the reality of what previously could only be imagined. Stone tools had to become proxies for ancient humans. And when found with extinct animals, they became the proxies for very ancient times. Imaginative fancy became hard-nosed fact.

Support came from the unlikely source of laws on the granting of patents. Throughout the 1850s John and his father-in-law John Dickinson had fought a legal challenge brought against them for the patent on their envelope-making machine. Their design of this highly profitable machine was challenged by Warren de la Rue (1815–89), a rival London-based papermaker and, like Dickinson, an FRS.[61] In 1857 the judgment, which broke John Dickinson and paved the way for the handover of the company to Evans, went against them. They had infringed de la Rue's intellectual property and were ordered 'to destroy or render useless' their envelope machines.[62]

As the historian of science Jenny Bulstrode points out, the precise language that is needed to defend a patent served John well when he came to make the case for the acceptance of stone implements. The language of patents deals in design and its precise and accurate description. De la Rue's case rested on a detailed description of what he called the 'peculiar character' of his machine. In 1857 the Evans-Dickinson envelope maker was compared to the following description in de la Rue's patent: it did not matter that their apparatus 'varied in form'. What mattered was whether it shared this individual defining 'peculiar character'. The ruling against them said that it did.[63]

And what was this 'peculiar character' on which the case hung? In the words of Robert Hawkins, who exactingly compared the two machines in January 1857, it was the mechanism by which 'the cam surface & its edge gives the final creasing under pressure'.[64] When digging through the legal archives, Bulstrode discovered that Evans had annotated Hawkins's opinion in pencil: 'No!', he wrote angrily. 'The folding is accomplished by the Edges.'[65] With so much at stake in terms of both reputation and profit she rightly concludes that this lesson impressed on John the importance of 'edges as tools, as products, and above all as points of discernment for technique and invention'.[66]

But what is the evidence? It is now a simple matter to tabulate the words and terms in these four pioneering papers. It turns out that the vocabulary they use is both similar and consistent. 'Edge', 'point', 'design', and 'shape' figure prominently (Table 3.1), supporting Bulstrode's argument.

What is striking also is their deliberate choice of neutral words such as 'implement', their favourite, and 'flake'.[67] The vocabulary of specific tool types is by contrast muted: 'axe' is hardly used. There is some mention of 'arrows', 'spearheads', and 'knives'—probably betraying Jacques's influence, just as 'weapon', the most popular description of the class of tool they thought the axes represented, points to Frere. Apart from 'weapon', their use of neutral language was deliberate. They were presenting facts, not interpretations.

Bulstrode goes further, arguing that Evans shaped deep human history in terms of a Victorian industrialist.[68] 'Manufacture' and 'workmanship' are used by both authors, but not overmuch. The language of the industrialist is there in the blue-paper letter of 25 May. The later polished axes, grouped generically under the title 'Celtic', are 'better adapted'[69] than the older ones from the Drift. The raw material, which for a paper manufacturer was a central concern, was less carefully selected to make unpolished rather than polished axes.[70] In *Archaeologia* Evans contrasts the pointed and oval shapes in terms of the 'remarkable feature' of the former and the 'ruling idea' of the latter.[71] The 'feature' of the former is that they were only adapted to cut or pierce at the pointed end, while the cutting edge of the polished, hafted 'Celtic' axes is quite the opposite. And for the latter, the 'idea' is the oval shape, more or less pointed.

The word search lends support to Bulstrode's idea that the two men resorted to making the case for the flints in the language of business and the demands of accuracy and precision applied to machinery. But that is only one suggestion. Another comes from the historian of science Nathan

Table 3.1 The vocabulary of flint tools, 1859

	Prestwich Abstract Proc. Roy. Soc 1859-60	Evans Abstract Proc. Soc. Ant. 1859-60	Prestwich Paper Transactions 1860	Evans Paper Archaeologia 1860
Number of pages	10	4	50	30
Edge	2	2	21	15
Point	2	2	15	14
Design	1	0	9	5
Shape	2	3	17	16
Struck	0	0	0	2
Manufacture/ workmanship	2	0	6	4
Wrought	0	1	0	1
Implement	19	23	113	97
Weapon	6	4	17	31
Tool/object	0	0	2	6
Flake	2	2	17	13
Knives	3	1	10	5
Spear- and lancehead	1	2	12	16
Axe	3	1	2	2

The vocabulary applied to flint implements in the abstracts and papers published in the *Transactions* and *Archaeologia* between 1859 and 1860. Even allowing for their different lengths and the topics covered, the similarity of usage points to a common understanding between Evans and Prestwich about the presentation of their case for human antiquity.

Schlanger. He argues that Evans also drew on the technical language of numismatics. John had published his first paper on pre-Roman British coinage in 1850, and it is still regarded as a classic.[72] In it he agreed with the geologist Gideon Mantell (1790–1852), who drew comparisons between the work of the palaeontologist and that of the numismatist. Fossils, Mantell wrote in 1854, were medals of creation.[73] It is not surprising that Evans's paper on 2 June started with his excursion into the Roman Cemetery at Saint-Acheul and the purchase of a coin.

But the vocabulary of numismatics is poorly represented in these four papers. Evans uses a coin word like 'wrought' only once—'flint-implements wrought by the hand of man'.[74] The only other numismatic word that he uses is 'struck', and that only twice: first to describe the coin struck at AMBIANUM (Amiens) and then to describe flakes 'struck' from flints.[75]

This seems insufficient to argue that prior to his own flint-knapping experiments he fell back on numismatics.

What Schlanger rightly draws attention to is that coins struck from a die do resemble flint implements, because in both cases great diversity is manufactured within an overarching general class—exactly the situation John had uncovered among the stone tools with their variable oval and pointed shapes and sizes. It is, Schlanger writes, 'as if each coin struck with the same set of dies was but a single point within a vast statistical cloud, scattered around a central, ideal, node'.[76] This point cloud of variation arises from the manufacturing process, the raw material, and the expertise of the artisan. In other words, Evans the numismatist expected his flints to vary substantially rather than narrowly conform to type. And when he came to consider time, his classic chart from 1850 showed how ancient British coins were derived from a stater, an ancient Greek coin, of Philip II of Macedonia (Figure 3.4). This was achieved by a process of imitations of imitations, a gradual evolution of new varieties but all still in the form of coins.[77] Many years later, Evans would draw a comparison between the evolution of coins and how natural selection worked in biology.[78]

Even though sorting out the geology of the Drift was a tricky business, Prestwich had it comparatively easy. He at least was working with well-established geological terms and concepts. Evans was not. And given his slight knowledge of flints in May 1859, his *Archaeologia* paper is an extraordinary achievement—extraordinary also because he had very little time to prepare it, squeezing the hours from three busy days before 2 June.

Illustrating the Claim

But there was another way for them to make their case—illustration.[79] Frere had set the standard of illustration with the highly detailed and aesthetically pleasing engravings by the artist Thomas Underwood who was draughtsman-in-ordinary to the Antiquaries. Drawn to scale and showing the weapons in profile and plan, these images seduce the eye (Figure 3.5).

A similarly high standard is present in the five implements illustrated in Evans's *Archaeologia* paper. He transferred his technical skill in designing machinery for Dickinson's to depicting flint implements, although he had Underwood's example to follow.[80] One is a flint flake, either a knife or arrowhead, from Menchecourt.[81] John was cautious about this piece: probably human manufacture he thinks, but nothing diagnostic to distinguish it

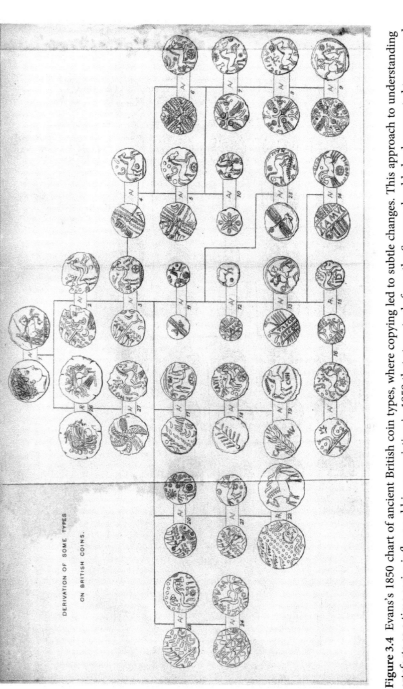

Figure 3.4 Evans's 1850 chart of ancient British coin types, where copying led to subtle changes. This approach to understanding artefacts as a time series influenced his appreciation in 1859 that stone tools from the Somme should also be expected to vary and change over time. The chart starts at the top and percolates downwards like a family tree.
Source: Evans 1849–1850.

Figure 3.5 Thomas Underwood's woodcuts, which accompanied John Frere's letter (Figure 3.2) about flint implements from Hoxne, published in 1800 in *Archaeologia*. The images were shown at full size.

Source: Frere 1800.

from the later 'Celtic' stone tools. He pinned his money on the three large implements he illustrated from Saint-Acheul, two varieties of the pointed and one oval example. He regarded the pointed forms as analogous to spear- or lanceheads tied, he conjectured, to a shaft.[82] And he added to them an object that his friend Augustus Franks unearthed in the collections of the British Museum. Found opposite Black Mary's Well, near Gray's Inn Lane in the City of London in 1679 by John Conyers, this triangular black flint was kept as a curiosity.[83] It turns up in John Bagford's letter of 1715 (he was one of the three friends who started the Antiquaries in the Bear tavern) about the antiquities of London. It next surfaces in the collection of the pre-eminent eighteenth-century collector Sir Hans Sloane.

There are many things about this black flint that are remarkable. It inspired curiosity. It was kept safe. It was transferred from collector to collector. It was found with the skeleton of an elephant. In a measured understatement John regarded its history as 'not a little singular'.[84] Obviously, he was delighted that Franks had found an older brother for Hoxne in the British Museum. The axe was part of Sloane's collection and transferred to the museum in 1753, when it became Britain's first publicly funded institution. For that reason alone the 'Black Mary' flint ought to be known as the 'people's axe'.

The oval-shaped implements were compared by John to some from burial mounds along the Mississippi. The *terrassiers* thought they were slingstones. He did not, because they are too large. Perhaps these are axes, or hatchets, as Boucher de Perthes suggested. But then John reined himself in: 'all this is conjecture'.[85]

The facts are the implements and drawings lying on the baize table at the Antiquaries. Let the fellows decide for themselves. In a statement which makes him sound like Robert Hawkins assessing the envelope machines, he expostulated: 'That they [the flints] really are implements fashioned by the hand of man, a single glance at a collection of them placed side by side, so as to show the analogy of form of the various specimens, would, I think, be sufficient to convince even the most sceptical.'[86] At the end of the day is it what your hands discover and your eyes confirm.

Eight implements from Hoxne, Abbeville, and Amiens are illustrated in Prestwich's *Transactions* paper. These include flakes as well as 'axes'. The implements, engraved by the family firm of Basire,[87] engravers to the Antiquaries, are shown at various scales, and while these are accurate illustrations in the tradition of Rigollot's axe, they lack the fine detail of those presented by the two antiquaries Frere and Evans.[88]

These arresting images counter any suggestion of forgery. To be absolutely sure, there is within Joseph's text a close-up of selected portions of two implements. These show a skin on the flint formed by a carbonate of lime, what is known as patina, and the fine tracery of root-like impressions, a result of being in the ground for a very long time. These are markers of antiquity and found commonly on the unworked gravels from which they were extracted.[89] Therefore, they could not be forgeries.

These two figures look even better in their original watercolours, which formed part of Prestwich's manuscript. Alongside them are sketches of the geological sections annotated by Joseph. The watercolours are probably the work of his sister Civil, who Grace records as making fair copies of his draft papers and undertaking the labour of organizing the references for his papers.[90] They are accurate summaries of the geology, with additions such as the well at T. Fréville's pit by the Cagny road and the trees at Hoxne that are then incorporated into Basire's engraving.[91] Such flourishes not only add a scale to the geology but a standpoint for the viewer, putting us in the picture. In 1859 photographs could not be reproduced.

The photographs were commented upon by both referees for Joseph's paper. The two chosen by the Royal Society ratchet up the meaning of 'peer review'. They were Sir Charles Lyell and Sir Roderick Murchison, the senior geologists of the day. Lyell sent in his two-page letter on 3 June. Publication was recommended. He had seen the photographs, and they bothered him; not because of their accuracy but how to convey what they showed. The photographs, he wrote, 'are very valuable and to a geologist convey a conviction of the undisturbed nature of the mass of gravel overlying the flint implement which nothing else short of visiting the place itself would do'.[92] The paper was passed on to Murchison who completed his recommendation on 1 July. He was also enthusiastic but wanted it returned to the author to correct minor mistakes, because it had been prepared 'in great haste', and it shows in the original manuscript. Murchison was also impressed by the photographs, which 'should certainly be represented, as they clearly exhibit an intermixture of materials which must have been brought about by a tumultuous and sudden operation'.[93] Despite these strong recommendations, the photographs were not converted into engravings.

In fact, they both went missing. The view of the pit at Saint-Acheul was once glued to page 24 of Joseph's manuscript (Figure 3.6). At the top of the page he described finding the artefact in Fréville's pit. He annotated the right-hand margin with the geological strata and added the orientation, West and East. When I put a copy of the photograph (Figure 2.11) back, it fitted snugly.[94] There is even a blob of glue in the top left-hand corner of the page that once held it in place. The photograph that accompanied the manuscript sent to the Royal Society is missing, as is the copy donated by Prestwich to the Antiquaries when Evans gave his talk on 2 June.[95]

One final aspect of the visual language is of interest. Four pages in Joseph's original manuscript are watercolour sketches of artefacts.[96] These have been cut out and pasted up as a page before going to Basire's for engraving (Figure 3.7). The axes are drawn with the thick end at the top and the thin, pointed end at the bottom of the page, much as though they were being held in the fist and used in a downward stabbing motion, living up to their other French name of *coups de poing*, 'punches'.[97] Wickham Flower published his flint implement in this way. The final versions of John and Joseph's papers reversed the orientation. This has the effect of indicating that change takes an upward direction, from a thick base to a thin tip (Figure 3.5). Time takes a positive direction.

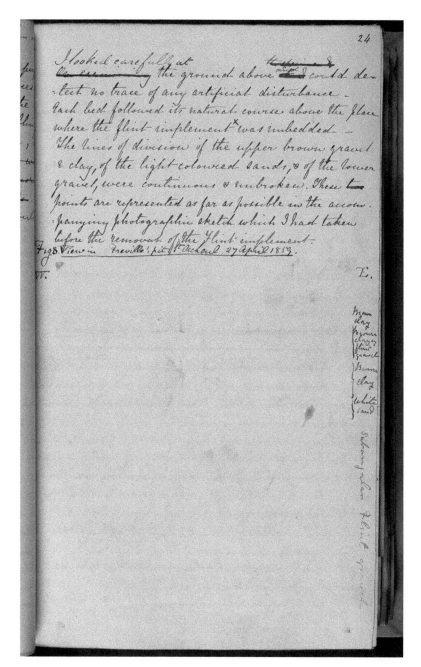

Figure 3.6 The missing photograph (see Figure 2.11) in Prestwich's Royal Society manuscript. The geological section, labelled West and East was annotated down the right-hand side, and he records where it was taken and when. This information did not make it into the printed version.

The handwritten text reads:

> I looked carefully at the ground ~~In examining~~ above ~~the specimen~~ and could detect no trace of any artificial disturbance. Each bed followed its natural course above the place where the flint implement X was imbedded. The lines of division of the upper brown gravel and clay, of the light coloured sands, and of the lower gravel, were continuous and unbroken. These ~~two~~ points are represented as far as possible in the accompanying photographic sketch which I had taken before the removal of the flint implement.

Fig. 8. View in Freville's pit, St Acheul, 27 April 1859 (Prestwich 1859a page 24)

His reference to a Fig.8 suggests that he once planned to turn the photograph into a drawing.

Source: Royal Society.

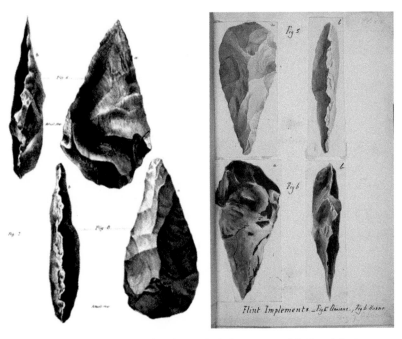

Figure 3.7 The watercolour sketches included in Prestwich's Royal Society manuscript that went to the engraver and the published images. Fig. 6 is from Hoxne and Fig. 8 from St Acheul. Notice how he follows Frere's convention and shows both the face and the profile of the pieces. The watercolours were probably painted by his sister Civil, who prepared his manuscripts. In the published version the orientation is reversed, as shown here, with the tips pointing upwards.

Source: Royal Society Prestwich 1859a:115.

1859 and War in Italy

And what of the other time revolutionaries travelling that month in Italy? Uncle Hugh, his niece Grace, and her maid Carolina Belloni arrived in Florence from Rome on 13 May. This was the trip of Grace's life. Since leaving the ice-cold Hôtel de Chépy the previous November, the small party had visited Paris, travelled from Marseilles to Genoa, which took them past Menton, like the two crinolined travelling companions in Augustus Egg's painting (Figure 2.7(a)), seen Vesuvius at dawn, discovered fossil hippopotami in the caves of Sicily, and taken in the sights of Rome. She met a famous society hostess, Mrs Mohl, in Paris, rescued Uncle Hugh's precious plaster casts of rhinos from the Naples customs house, and engaged with Italian and expat English society. Her journals are full of excitement for everything from gardens to dances to landscapes and ancient ruins, even bone caves.

The northern Italian states they travelled towards were preparing for war. On 23 April the Austrian emperor had sent an ultimatum to Piedmont's prime minister, Count Cavour, demanding that his forces withdraw from their western border, the River Ticino in Lombardy. Cavour was given three days and then Austria would attack.[98]

Cavour immediately telegraphed the ultimatum to his ally, Napoleon III, in Paris. This was exactly what their military and diplomatic manoeuvres had aimed to achieve. The Austrian Emperor Franz-Josef had taken the bait and in threatening to attack delivered them a public relations coup. Napoleon immediately gave the instruction to send French troops to Italy under the banner of liberating Italy from the imperial yoke of the Habsburgs. He had 120,000 men stationed in Lyons, Toulon, and Marseilles, and these now embarked for Genoa. The Imperial Guard entrained in Paris while other forces were sent across the Alps and then by train to the Po plain, where the rival cities of Turin, in Piedmont, and Milan, in Lombardy, were facing off. In the two weeks to 12 May over 70,000 troops arrived in Genoa. *Le Monde illustré* had a patriotic field day. Every issue carried pictures of troops marching to war, cheered on enthusiastically by grateful civilians.

The Austrians' bluff was called, and war broke out when the ultimatum expired on 26 April. Planning for this war had been meticulous, and the coalition of France and Piedmont made use of the astonishing speed of the railways to move troops from the ports to the Ticino. It would be a truly modern war, broadcast by telegraph, speeded by rail, and for the first time photographed stereoscopically.[99]

The campaign moved to the offensive the day Uncle Hugh arrived with Grace in peaceful Tuscany. Known as the Risorgimento, the Second War of Italian Unification, or the Franco-Austrian War, it would prove decisive in leading to the eventual unification of Italy's fragmented city states into a modern nation state with its capital in Rome. But in 1859 that was still in the future. The view from England was anxious. *The Times* saw it as another example of French mischief-making. Ferdinand de Lesseps's pickaxe had broken ground for the Suez Canal at Port Said on 25 April, the day before this phase of the Risorgimento began.[100] The concern now was over Napoleon III's further imperial ambitions in Europe, and the spectre of his hated uncle was never far away. The editorial on 26 May, after the first battle at Montebello six days before, offered Napoleon III this avuncular advice: 'make a short, sharp, decisive campaign, or expect to hear murmurs in [your] camp and to read of disaffection in [your] capital.'[101] John Bull felt duped. Two months before, 'France...was unanimous for peace. Not a domino-player looked up from his game to utter a profane prayer for the return of the days of the First Empire. But the war came.'[102] Quick and dirty was the best Britain could hope for. Otherwise, she might get sucked in to a long war of attrition.

Giant forces were mobilizing while John and Joseph were visiting Abbeville and writing their papers, and Uncle Hugh and Grace were cataloguing fossils and travelling to Tuscany. They may have had anxieties, but none is recorded. All four are only loosely folded into this headline historical event, cushioned from it by being so wrapped up in their own immediate focus that could have unravelled at any time. Any rupture in relations between England and France would disrupt Prestwich's wine business and make importing continental rags, essential for Evans's papermaking, extremely difficult. Strikingly, there is no hint that Joseph thought the anxious mood in London might prejudice their scientific case because the flint implements had been found in France. The principle of *Nullius in verba* held sway.

A bubble enclosed the visitors in Italy. The day after the Austrian ultimatum was delivered, Grace wrote in her diary, 'To St Peter's to see the Benediction on Easter Sunday. The illumination was very fine in the evening, the Cross bursting at once into a star of light. It looked very gorgeous from Monte Pincio,'[103] and a month later in Tuscany after the battle of Montebello had been fought, 'The Pitti Palace, the Uffizi, the Duomo, the churches, the Cascino Gardens, the Boboli, the expedition to Fiesole, the beauty of the situation,—all made me enjoy Florence more than any other city of Italy.'[104] But the war did concern other expats living the Italian dream. One of these was Mary Somerville, who invited Uncle Hugh's party for tea at her house in the via del Mandorlo. This was on 29 May. Mary was a trailblazing woman and an energetic 79. She had escaped a bad marriage to Captain Samuel Greig when he died in 1807, after which she inherited his fortune. Five years later she married her cousin Dr William Somerville.

Mary was an indefatigable champion of a woman's right to an education, an idea vehemently opposed by Greig but sympathetically supported by Somerville.[105] She was a brilliant mathematician, and two of her most famous books are *On the Connection of the Physical Sciences* (1834) and *Physical Geography* (1848). Her range of interests is breathtaking. She published on astronomy and molecular and microscopic science, and was just the sort of scientist who, if male, would have been elected instantly to the Royal Society. Life in Florence was far more congenial to her many talents, and there were always interesting people, like Falconer, a fellow Scot, dropping by to escape the northern cold.

The Falconer party had grown. It now included another niece and his brother, who had spent rather too long in Australia and lost the art of polite conversation. Mary was most interested in Uncle Hugh, and he regaled her with the fossil finds that he and Grace had found in Sicily while also telling her about Brixham and his hopes for the Somme. Grace was captivated, not just by Mary's ability to engage scientifically with Uncle Hugh on a wide range of topics, but also by how she dealt so charmingly with the awkward uncle and his anecdotes of life in the bush.[106] She started to look around the room for Mary's work basket. But despite her reputation as a scientist who was an accomplished needlewoman, it was absent. Uncle Hugh had told the story from an earlier visit, when he found the author of *The Mechanisms of the Heavens* (1830) darning the family's stockings. Their visit is described in a letter Mary wrote afterwards to her son. She had spent the day making bandages for the soldiers fighting for Italian independence. Then she went on to recount Uncle Hugh's discoveries of 'implements formed by man, mixed with the bones of prehistoric animals, in a cave so hermetically shut up, that not a doubt is left of a race of men having lived at a period far anterior to that assigned as the origin of mankind. Similar discoveries' she went on, 'have recently been made elsewhere.'[107]

Grace did get much closer to the war. Falconer remained in Italy, and she returned with another English party via Genoa, where they lost their pass-ports, and then on to Turin, crossing the Alps at Mont Cenis, and so to Geneva, Paris, and London where she arrived on 29 June. She saw the massive fortifications at Alessandria and many troops, horses, and cannon. Because of the delay caused by the lost passports, her party was on the last train to Turin on the day of the battle of Magenta, a battle which resulted in a less than conclusive victory for the French-Piedmont army. Reuters telegraphed the news around Europe, and it was on the front pages the next day. The French telegram announced, 'A great victory. 5,000 prisoners. The enemy lost besides 15,000 men,' while the Austrians played down their setback: 'The combat was undecided...a further fight is expected.'[108] Grace was only 25 miles from the battle on 4 June. She wrote in her diary:

> there was a suspicion of risk about the journey that in our utter ignorance of warfare gave it a zest to some of us—though not to all...The Piedmontese and French troops...lost 4000 men killed and wounded. One shudders in reading the bare record. Could we wonder that the railway officials looked sick at heart and went sadly about their work that day?[109]

While the carnage at Magenta occupied the nations of Europe, John was amending for publication in the *Transactions* the letter on blue paper he had written to Prestwich on 25 May. What their discoveries had created, he now wrote, was a 'neutral territory between Palaeontology and Archaeology'. Here, 'a wide field is opened for investigation, which must eventually lead to a great extension of our knowledge of the history of primeval Man'.[110] In the letter following her tea party for Grace and Uncle Hugh, Mary Somerville, ahead of the game as usual, wrote of 'prehistoric animals'. Prehistory would be the wide field of knowledge that changed the shape of human history. And what better that a Scotswoman, Somerville, should be writing about a visit from Falconer's Scottish niece, McCall, and use the word prehistory, coined in 1851 by the Edinburgh antiquary Daniel Wilson.[111]

Notes

1. Evans, 1949: 122.
2. Evans 1943: 102, Letter from John Evans to his fiancée Fanny Phelps in Madeira, 5 May 1859. On 2 May Prestwich wrote in a postscript to one of the invitees, George Waterhouse, 'I would ask the great Professor if I thought I could & he would come,' Most likely this was Sir Charles Lyell, but he was not present. George Waterhouse was a Keeper in the British Museum and a leading naturalist specializing in descriptions of mammals, particularly marsupials and rodents. He

published the collection of insects that Darwin had brought back from the voyage of the Beagle, a voyage Waterhouse was invited to join but declined. Letter belonging to Jeremy M. Norman.

3. The purpose of the London Clay Club was to illustrate, with scientific accuracy, Britain's fossil flora and fauna, work that continued with the Palaeontographical Society. See http://www.palaeosoc.org/site/home/, accessed 27 July 2020.

4. Grace Prestwich (1899: vii) thanked Dr Henry Woodward (1832–1921), another member of the London Clay Club, for the photograph but gave no date.

5. Published between 1848 and 1856.

6. Failing health prevented him from completing his richly illustrated volumes on the Eocene Mollusca of Britain, the Club's original project.

7. His obituary appeared in 1876: see https://www.cambridge.org/core/services/aop-cambridge-core/content/view/S0016756800160625, accessed 27 July 2020.

8. Dickens wrote entertainingly about his visit to Wroxeter, *Rome and Turnips* in *All the Year Round* 14 May 1859, pp. 53–9. He asked why money was spent digging up Nineveh, when for a smaller sum something significant could be discovered about British history.

9. Evans 1943: 102.

10. A comparative newcomer, the Geological Society was founded in 1807.

11. https://royalsociety.org/about-us/history/, accessed 17 November 2016. It was inaugurated in 1660 with a lecture from the architect Christopher Wren to an 'invisible college' of natural philosophers and physicians. A Royal Charter soon followed. The Royal Society was lampooned by Jonathan Swift in *Gulliver's Travels* (1726) as the kingdom of Balnibari where impractical experiments are undertaken.

12. Evans 1956: 36. The three friends were Humfrey Wanley, John Talman, and John Bagford. The handwritten minutes survive (Sweet 2007: 56). With its emphasis on British antiquities and history, the Antiquaries set themselves as a balance to the British Museum, which at this time was a shrine to classical civilizations, the epitome of the Enlightenment.

13. 'Agreed that while we meet at a Tavern, no person shall be oblig'd to pay for more than he shall call for' (Evans 1956: 36).

14. The Royal Society moved to Burlington House, Piccadilly, alongside the Royal Academy in 1857. The Antiquaries followed them seventeen years later.

15. Sweet 2007.

16. Evans 2007: 186.

17. https://nttreasurehunt.files.wordpress.com/2014/11/cruikshank.jpg, accessed 27 July 2020.

18. Prestwich was a fellow of neither the Antiquaries nor the Numismatic, but he joined the Ethnographicals in 1869 as it was transforming itself into the Anthropological Institute (see Chapter 6).

19. Secord 2007: 29.

20. van Riper 1993: 104–13.
21. Prestwich 1859–60: 51.
22. Evans 1868: 93.
23. Prestwich 1864b: 213.
24. Gruber 1965: 385. The Brixham Committee set up in 1858 included Falconer and Lyell as well as Richard Owen and Godwin-Austen, with the work falling to William Pengelly and Prestwich.
25. The pressure Buckland exerted on MacEnery, who died in 1841, is widely believed to have forced him to change his testimony (Grayson 1983: 74; White and Pettitt 2009; Pettitt and White 2011).
26. Prestwich 1860: 278.
27. See Sommer (2004) for the motives behind Buckland's interpretation and suppression of evidence.
28. Prestwich 1860: 303–4.
29. Prestwich 1859–60: 51. Falconer had also written at length to the Geological Society on 10 May 1858 about the potential of Brixham as a bone cave (The letter is reproduced in Prestwich 1873b: 471–5). In this earlier letter he made no mention of stone tools.
30. G. A. Prestwich 1899: 123.
31. Evans 1868: 94.
32. G. A. Prestwich 1899: 125. The letter is dated 18 May 1859.
33. Prestwich 1860: 305. Frere died in 1807.
34. Prestwich 1860: 306.
35. These stone implements are also known by the French *coups de poings*—punches (Burkitt 1923: 92).
36. Evans in Prestwich 1860: 312.
37. Babbage 1859–60: 69. He based his opinion on observations of how gun flints were made, an experimental route that Evans would follow.
38. Prestwich 1859–60: 58–9. This category was in the blue-paper letter that Prestwich had left behind.
39. Evans 1943: 103.
40. They went on 16 April 1860 (Hurel and Coye 2011).
41. Owen (2013: 5), quoting from notes by Lubbock reproduced in Hutchinson (1914: 23–4).
42. Owen 2013: 19. The paper, 'On Certain Sensory Organs in Insects, Hitherto Undescribed', was by John Braxton Hicks (1823–97), an obstetrician and biologist.
43. *Darwin Correspondence*, Lubbock to Darwin, 8 February 1859. In his letter to Darwin, Lubbock is concerned with the continuing saga of the poor health of the great naturalist and his daughter, and a disfiguring accident to the hand of his brother, Beaumont Lubbock.
44. *Darwin Correspondence*, Darwin to Lubbock, 6 February 1859.

45. Lubbock 1859.
46. *Darwin Correspondence*, Lubbock to Darwin, 15 March 1859.
47. Evans 1943: 103.
48. His name does not appear in the Antiquaries visitors' book for that meeting. Pouchet (1860 in de Brassac 1999: 300) states that Prestwich's second visit was on 28 May. The other two geologists were Robert Godwin-Austen and Robert Mylne. Godwin-Austen was a student of Buckland's at Oxford. He excavated at Kent's Cavern and was a member of the Brixham Cave Committee. He did not hold Buckland's fundamentalist views. Robert Mylne published the first geological map of London in 1856. Prestwich knew him through this work, which was an essential tool in the drive to improve public health by understanding the capital's water supply (Evans 1949: 122; G. A. Prestwich 1899: 124; Hurel and Coye 2011: 17).
49. Joseph recalls that, besides human antiquity, Flower's great passion was to determine the origin of the pebble beds of the Addington Hills, Croydon: 'He broke up many thousands of them in search of the small fossils they occasionally contain. Unfortunately, the results of this long investigation have never been published' (Prestwich 1873a: 430).
50. Flower 1860, engraving by S. J. Mackie.
51. Evans 1860a: 282. Prestwich (1860: 280) includes this observation as a footnote.
52. Prestwich 1860: 294.
53. Prestwich 1860: 294.
54. Prestwich 1860: 295.
55. Prestwich 1860: 295.
56. Evans 1860a: 289.
57. Ruskin 1853: 167.
58. Ruskin 1853: 167.
59. Ruskin 1853: 172.
60. Penwarden and Stanyon 2008; Stewartby 2008; King 2008; de Jersey 2008.
61. De la Rue was renowned for his early photographs of the moon and solar eclipses.
62. Bulstrode 2016: 5–6.
63. Bulstrode 2016: 6.
64. Bulstrode 2016: 7.
65. Bulstrode 2016: 8.
66. Bulstrode 2016: 8. It is worth also remembering Ruskin's stricture about Venetian glass and edges:

> Now you cannot have the finish and the varied form too. If the workman
> is thinking about his edges, he cannot be thinking of his design; if of his
> design, he cannot think of his edges. Choose whether you will pay for the

lovely form or the perfect finish, and choose at the same moment whether you will make the worker a man or a grindstone. (1853: 168)

67. Among other neutral words, Evans uses 'tool' twice in his blue-paper letter and 'object' only appears in *Archaeologia*.
68. Bulstrode 2016: 4.
69. Evans in Prestwich 1860: 311.
70. Evans 1860a: 288.
71. Evans 1860a: 291.
72. de Jersey 2008: 156–8; Stewartby 2008: 201–2.
73. Schlanger 2010: 344. Mantell's book was published posthumously, and its full title is *The Medals of Creation or, First Lessons in Geology, and the Study of Organic Remains*.
74. Evans 1860a: 280.
75. Evans 1860a: 285 and 289.
76. Schlanger 2011: 473.
77. de Jersey 2008: 156.
78. Evans 1875: 487.
79. Smiles 2007: 123; Kaniari 2008.
80. Kaniari (2008: 279 and n. 58) draws attention to Evans's technical expertise in designing machines and how this was transferred to his flint drawings. The dry and hard edge style he employed links to the needs of industrial design.
81. Evans 1860a: 290 Evans's approach to illustration leaves no doubts about intentional design. At this stage he is less interested in the extent of variation. That point cloud will come later (see Chapter 5). The artefact from Menchecourt was the first flint implement to be published since Hoxne. It appeared in the magazine *Once a Week* on 2 July 1859 (see Chapter 4).
82. Evans 1860a: 291.
83. The history of this remarkable axe can be found in White, Ashton, and Lewis (2018). Today it is commemorated by an office block at 277A Gray's Inn Road named Hand Axe Yard.
84. Evans 1860a: 302.
85. Evans 1860a: 292–3.
86. Evans 1860a: 288.
87. William Blake was apprenticed to this firm of engravers for seven years.
88. Kaniari 2008; Roberts and Barton 2008. John continued to develop his visual language, which culminated in his *Ancient Stone Implements* published in 1872.
89. Prestwich 1860: 297, Figs 8 and 9.
90. G. A. Prestwich 1899: 102.
91. Prestwich 1860: 289, Figs 5 and 305, Fig 12.
92. Lyell, letter, 3 July 1859, Royal Society MSS RR 4 207 1.

93. Murchison, letter, 1 July 1859, Royal Society MSS RR4 208 1.
94. There is a space of 16.5 × 21 cms on the page into which the Pinsard photograph in Amiens just fits.
95. Evans 1860b: 333.
96. Royal Society Library MSS RR 4 209 4 1860.
97. Burkitt 1923: 92.
98. Schneid 2012.
99. The Crimean War had also been photographed, but the role of the railways was negligible. For Franco-Austrian War photographs, see https://vintagephotosjohn son.com/2012/02/18/combat-photography-during-the-franco-austrian-war-of-1859/, accessed 31 September 2020.
100. See note 110 in Chapter 1.
101. *The Times*, 26 May 1859, p. 8.
102. *The Times*, 26 May 1859, p. 8.
103. Milne 1901: 31.
104. Milne 1901: 31. The entry is dated 23 May.
105. Secord 2007: 33; Lightman 2007.
106. Milne 1901: 126.
107. Milne 1901: 130. The letter was dated the same day, 29 May, to her son Woronzow Greig.
108. *The Times*, 6 June 1859, p. 7.
109. Milne 1901: 32.
110. Evans, in Prestwich 1860: 312; see also Gamble and Moutsiou (2011: 50) for the changes Evans made to his original letter.
111. Wilson 1851.

4

Reception

The Year 1859–60

A new human history of unfathomable antiquity was now possible, its timescale no longer built on the evidence of the fragile written word but on indestructible implements of stone encased in a geological setting. Until the time revolution of 1859 human histories were like the hand game where paper always defeats rock by wrapping it. Now, thanks to Boucher de Perthes, Prestwich, and Evans, stone had become the cutting edge of deep history.

How their time revolution was received would depend on how they answered three pressing questions: when did human history now begin? Were these stone tools authentic human handiwork or due instead to natural causes? And lastly there was the necessarily speculative, but nonetheless contentious issue of exactly how old *was* old.

Both the time revolution and Darwin's biological upheaval are often presented as a clash between new science and established religion, faith and evolutionary theory, as epitomized by Huxley debating with Bishop Wilberforce at Oxford in June 1860. But for human antiquity this is a simplification. Anglican clerics held a wide range of opinions about the age of humankind and the evidence which supported it. But their prominent voices disguised a paradox. While the country regarded itself as a Christian one, almost half who attended religious services did not worship in its established churches.[1] This helps explain the violent reaction to a book with the seemingly innocuous title *Essays and Reviews*, published in March 1860 by six clergymen and one Egyptologist with an interest in geology, Charles Goodwin (1817–78). The seven essays contained little that was new either in theology or in biblical studies. What called down the fury of the Church of England was *who* was proposing that the Bible should be read like any other book and that reconciling geology and Genesis should be abandoned in favour of the former.[2] Two of the clergymen were tried and found guilty of heresy by the Church's Court of Arches. The book was officially condemned in 1864,[3] a fate which *The Origin* avoided. Indeed, a study of the column inches devoted to comments and rejoinders in the newspapers about *The Origin* and *Essays and Reviews* reveals the latter as outright winner.[4]

The official keepers of religion were, therefore, much more concerned by serious matters of internal discipline than by fresh geological evidence for human antiquity.[5] But what about their competition, the Evangelicals, whose congregations were growing? What was their reaction to challenges to Holy Scripture and the 6,000-year timescale for the Creation, as laid down in Genesis? Their response was varied. Many rejected deep human history as unbelievable, and they still do. The Bible, they maintain, is not to be approached historically but always believingly. Some, however, found it possible to accommodate the evidence of geological deep time with belief in a Creation that took six days. This involved stretching the length of a day. Others were actually heartened by the demonstration that humans lived before Adam.[6]

The wider reception of Joseph and John's time revolution was never a simple matter that pitted the word of God against evidence from the hand of Man. And while the established Church pursued its internal heretics, there were other issues swirling around in 1859 that shaped the argument for human antiquity and which Prestwich and Evans hoped would gain acceptance beyond their circle of immediate friends.

The year is notable for the clarity with which the tension between individual freedom and the constraints imposed by society was articulated. Such a debate has always forged a connection between individuals and their collective history. The argument is there in *Essays and Reviews*, with its challenge, by promoting individual interpretation, to the Church's authority over the message of the Bible.[7] At the beginning of 1859 John Stuart Mill (1806–73) had propelled discussion of the issue to the forefront with his influential *On Liberty*. It was well received by another public intellectual, Herbert Spencer. He wrote to Mill on 17 February congratulating him on his treatment of 'the claims of the individual versus those of society' and bemoaning 'that mania for meddling which has been the curse of recent legislation'.[8]

The eternal tension between self-interest and social contracts ran throughout the year, shaping private and social discourse. Political choices had to be made by the tiny electorate permitted to vote in 1859.[9] Parliamentary elections were held in April and May. The Tories fell and the first Liberal government, a coalition of minority parties, succeeded them in June. Free trade was now on the agenda. Against this background the historian Henry Buckle enthusiastically reviewed *On Liberty* in May. He praised Mill for championing the individual against the forces of the state and public opinion, which Buckle regarded as the 'voice of mediocrity'.[10] What Buckle proposed was a philosophical history freed from divine intervention that involved neither predestination nor the gift of rampant free will. He replaced these explanations for change with a basic historical dialectic: the modification of humans by nature and vice versa.[11] History was taken away from God and put into the hands of men like Lubbock, Evans, and Prestwich. They were also among the few males who, like Buckle, could vote.

The tension between society and the individual took many opposing forms, culminating on 24 November with publications by Darwin and Smiles. Their two very different books provided the factions wrestling over liberty and responsibility with case studies to quarry for their cause. Darwin showed in *The Origin* the importance of the population as the biological crucible for change; by contrast, Smiles championed in *Self-Help* the power of the well-motivated individual to transform the world. Darwin's evolutionary journey was blind, with no goal or purpose driving it. This was not the case with Smiles's movers and shakers of the nineteenth century, men like Prestwich, Evans, and Lubbock. They were driving the engine of progress.

Spencer saw no such tensions when in 1857 he put forward in his essay *Progress: Its Law and Cause* a principle that sought to unify biology, culture, and history. He proposed that 'it is settled beyond dispute that organic progress consists in a change from the homogeneous to the heterogeneous', or, to put it more pithily, what starts out simple always becomes complex. Here was a sublimely confident summation of a prosperous decade, widely regarded as the Victorian heyday.[12] Spencer's unifying theory was breathtaking in scope:

> Whether it be in the development of the earth, in the development of the life upon its surface, the development of society, of government, of manufactures, of commerce, of language, literature, science, art, this same evolution of the simple into the complex, through a process of continuous differentiation holds throughout.[13]

Beneath the canvas of such a big tent, Spencer easily accommodated the arguments in *On Liberty* in January, the idea of great human antiquity in April, and by the end of the year the principle of descent with modification through another mechanism effecting change, natural selection.[14] Spencer's all-embracing mantra that simple precedes complex was seductive for men like Prestwich and Evans who wanted facts not theory and a clear direction to human history, something that was not possible if they read Darwin correctly.[15]

These tensions and the ingenious philosophy of Spencer shaped the social and scientific attitudes of the audiences that reviewed their case for a deep human history. Religious belief was secondary to individualism in this age of science and relentless innovation. A better understanding of why things changed was simply more important to privileged, franchised men like Evans and Prestwich than dog-eared theology.[16] This tension between the individual and the population from which they were drawn—the state and society—framed the reception of John and Joseph's discovery and presentation of a new, deep history.

Three Ages When History Began

The history that arose from the 1859 time revolution broke new ground. What Evans and Prestwich had demonstrated was novel because of the geological time-depth they recorded in such painstaking detail. By validating Boucher de Perthes's claim that chipped stone tools pointed to an antediluvian age for humans, they had taken human history back beyond Noah's Flood and before Adam and Eve. Moreover, they were about to add another strata beneath the prehistory devised by the Northern Antiquaries of Denmark and Sweden.

To appreciate when European history began in 1859 it is necessary to trace the Roman wall that once divided the continent. This was the fortified *Limes Germanicus*: the frontier between the civilized world of Rome and the eternally troublesome barbarians who lived beyond. Running for over 550 kms, the *Limes* used the Rhine as a defensive barrier, while taking in some of the territories immediately to its east. About a quarter of modern Germany lies within the empire, while Denmark and Sweden are completely outside. A smaller brother of the *Limes*, Hadrian's Wall, similarly divided Roman Britannia from Barbarian Caledonia, Scotland.

Answering the question 'When did history begin?' was straightforward for an antiquary living in a nineteenth-century nation state that happened to have once been under Roman protection. History started with the arrival of the civilizing process. Once named by classical authors such as Julius Caesar, large areas of Europe swam into historical focus, like a photographic plate in a tray of chemical developer. Monuments such as forts, aqueducts, town walls, and occasionally public buildings, along with rural villas and roads marked a significant page in national histories, a turn for the better.

Historians of the nineteenth century, immersed in the imperial ambitions of their age, rarely looked any deeper. The statesman Thomas Macaulay, who died in 1859, opened his optimistic account of the *History of England from the Accession of James II* with a couple of pages on Britain under the Romans. Even so he did not particularly like what he saw: 'Subjugated by the Roman arms... she received only a faint tincture of Roman arts and letters' so that 'nothing in the early existence of Britain indicated the greatness she was destined to attain'.[17] But even worse was what came before the Romans: 'Her inhabitants ... were little superior to the natives of the Sandwich Isles [Hawaii]'.[18]

Others saw the starting point differently: Edward Augustus Freeman (1823–92), whose daughter Margaret married John Evans's eldest son Arthur in 1878,[19] regarded the Norman Conquest not as the beginning of the English nation but the turning point in its history.[20] He used historical facts to bury any dissenting voices to his thesis that assimilation of the Normans, not conquest of the English, was the correct reading of the evidence. British identity emerged from this historical process.[21]

History for the wildly eccentric and rather unpleasant Freeman consisted of 'memorable facts',[22] an approach lampooned subsequently by Sellar and Yeatman in *1066 and All That,* the precursor of many a horrible history.[23] Freeman mentioned the Romans in his massive *History of the Norman Conquest of England.* But they were about as important to him as they were to Macaulay. The literate empire of Rome provided an acceptable point of origin. However, the turning points in British history came either in 1066 for Freeman or 1688 for Macaulay, with the Glorious Revolution. These were the true beginnings of the 1850s gilded Age of Equipoise.

English historians lived in a country that historically lay within the protection of the *Limes Germanicus*. Their histories might start at different times but always with a fully literate society. What happened outside that imperial embrace suggested a different answer to the question of when history began. Nowhere in Europe was this more strongly felt than in Scandinavia and, in particular, Denmark and Sweden. Without a Roman presence, their antiquaries came up with something different. Early in the nineteenth century the agricultural revolution bit deeper into the soil, yielding a cornucopia of artefact and human remains. Northern

Antiquaries began to use this material evidence to argue for what they called a prehistoric beginning to their history.[24] Gradually collections accumulated as curiosities in newly founded museums. Then, in 1836, at the Danish National Museum in Copenhagen Christian Jürgensen Thomsen (1788–1865), a contemporary of Boucher de Perthes, organized those in his care into three ages. These were the Stone, Bronze, and Iron Ages, their historical order confirmed by archaeological excavations. Danish history now began in a remote, Prehistoric Stone Age. The deep roots this implied were significant. The patriot and archaeologist Jens Worsaae declared in 1849, the year Denmark became a constitutional monarchy, that the remains of antiquity in the prehistoric monuments of Denmark:

> constantly recall to our recollection, that our forefathers lived in this country, from time immemorial, a free and independent people, and so call on us to defend our territories with energy, that no foreigner may ever rule over that soil, which contains the bones of our ancestors, and with which our most sacred and reverential recollections are associated.[25]

This sentiment was widespread: 'No man', wrote the English antiquary John Mitchell Kemble (1807–57) in the year he died, 'values higher than myself that noble spirit which makes us look with love upon the records of our own ancestors, and of our own land.'[26]

The earliest use of the term prehistory (Forhistorisk in Danish) was in 1834 and commonplace in Scandinavia by 1849.[27] In Britain it was first used by Daniel Wilson (1816–92), an antiquary living to the north of Hadrian's Wall in Edinburgh. His book *The Archaeology and Prehistoric Annals of Scotland* appeared in 1851.

Europe's Roman wall divided opinion on when history began. To the west, as in England, France, and Spain, it relied on Roman texts and literature. The classical worlds of the Mediterranean and the biblical Near East provided the inspiration for history. However, to the east and north of the wall an earlier phase existed, as shown by the evidence of implements. These could be organized into stages that showed a degree of development as metal replaced stone tools. Very soon these technological stages—Stone, Bronze, and Iron—became historical phases, with the earliest representing hunters and the latter two pastoralists and farmers.[28] Without any prompting from Spencer the empirical stages enshrined in Thomsen's museum cases evolved 'naturally' from simple to complex histories. And history in three parts has dominated the ordering and interpretation of archaeological evidence ever since.[29]

But the three-age system of Northern Antiquaries such as Worsaae and Thomsen lacked one important element, unpolished stone axes. These were Rigollot's and Boucher de Perthes's *haches antédiluviennes*, which Evans and Prestwich had presented in May and June as proof of great human antiquity. The reason they were unknown throughout Scandinavia, and remain so, is due to ice. Successive ice sheets swept down from the high Fenno-Scandinavian shield to cover Sweden and Denmark. In the process the superficial deposits with evidence for the earliest human occupation in these northerly latitudes were scraped away. When Thomsen came to classify the Danish collections, he had only polished axes from the Holocene to put into his Stone Age. The older Pleistocene axes did survive in eastern England, at sites like Hoxne, with its fighting stones, due to a complex geological history at the margins of the ice sheets. South of the River Thames, another rich area for hand-axe discoveries, was never glaciated, and this was also the case for the *langues de chat* from the Somme. By contrast, northern England, most of Wales, and all of Scotland and Ireland were the seat of major ice sheets which had ground away the earliest evidence for human occupation.

English antiquaries knew about Thomsen's three ages. His *Guide to Northern Archaeology* was published in English in 1848, but its reception was slow. In 1999, when the archaeologist Michael Morse was researching the adoption of Thomsen's system in England, he found that the pages in Oxford's Bodleian Library were uncut. No one had read it in 150 years.[30]

The antiquary John Kemble acknowledged in 1857 that the three ages might have some foundation in historical fact and considerable practical value when it came to arranging a national museum. But Thomsen's ages did not enshrine a universal history. The Stone Age, Kemble declared, did not stop with the arrival of metals.[31] In his opinion it was more important to understand local and national development rather than use the same antiquities to support a global historical scheme, as proposed by a northern antiquary such as Worsaae. However, he shared Worsaae's nationalistic fervour when it came to the power of objects. British antiquities for British people. Sovereign histories for a sovereign people. Let the record speak for indigenous values, rather than impose a universal history of three ages through which every nation must pass.

John Evans agreed. After that day in Amiens he devoted himself to the compilation of data. His three great volumes on *Ancient Stone Implements*,

Weapons and Ornaments (1872), *Ancient Bronze Implements, Weapons and Ornaments* (1881), and, for the Iron Age, *Coins of the Ancient Britons* (1890) might seem to follow the spirit of Thomsen's three ages. However, he put in the hard labour for a deep British history for a British audience. He had less interest in a world history as pursued either by his close friend John Lubbock, *Pre-Historic Times* (1865) or by the American social theorist Lewis Henry Morgan (1818–81) in his book *Ancient Society* (1877).[32]

He did, however, give one small inkling of the universal history favoured by the Northern Antiquaries with their three prehistoric stages. Not only did Evans claim that unpolished flint axes were as ancient as the Drift and its extinct animals, but he also used their position throughout the Somme's quarry pits to argue a basic evolutionary scheme, one measured by manufacture, and where in true Spencerian fashion things got progressively better with time, moving from rudely made to better made. Evans's 'history of primeval Man'[33] compared the flints of the Drift with the more recent 'Stone period' of the Northern Antiquaries.[34] What he did was take a vertical face of time, the Somme stratigraphy, and turned it into a timeline of human history. Both levels in the Somme, the Drift, and the misnamed 'Celtic'[35] had simple stone tools but the lower was cruder in workmanship than the upper. This, John claimed, pointed to a higher degree of civilization among the later.[36] Time's arrow shot ever upwards in human history.

But where might the interested see the evidence for this new, deep history? The British Museum, repository of the national collection, did make limited use of Thomsen's three ages. This was set out in a case-by-case guidebook to its galleries, published in 1859.[37] The pagination says a good deal about the relative importance of different parts of the collection. As might be expected for a country once protected by the *Limes*, the guide is dominated (48 pages out of 104) by sculpture and antiquities from the classical civilizations and the Near Eastern empires. A further forty-eight pages described the Natural History galleries.[38] That left just two pages for the rest of the world in the Ethnographic Room and five for the British collections. Here the visitor would find forty-two cases, just over a page in the guidebook, containing antiquities that predated the Romans. These were organized according to Thomsen's scheme to show when the three different materials—stone, bronze, and iron—were introduced to the British Isles. The oldest objects were polished stone axes in the first four cases of the room.

Earlier material now needed to be added. In May, Evans's friend Augustus Franks, one of the British Museum's most influential curators,[39] had unearthed from the collection an axe similar to those from the Somme, the 'people's' axe found at Gray's Inn, London in 1679. The Hoxne discoveries of John Frere in 1797 could now be added to this new case,[40] along with a steady stream of discoveries from the river gravels of southern England.

Were the Stone Tools Really Stone Tools?

None of this would matter if the stone tools were natural rather than artificial. After John's talk to the Antiquaries on 2 June, they awaited the reception of their case. First off the blocks was *The Athenaeum*, the must-read broadsheet of the serious-minded, which covered, in digest form, the business of learned societies, gossip, art, and books. John's abstract was published on 11 June.[41] It was then picked up by regional and local papers such as the *Hereford Journal* on 22 June, the *Caledonian Mercury* on 24 June, and *The Gentleman's Magazine* in early July.[42] Prestwich's Royal Society paper received similar syndication. Letters soon followed.

The first came from the 'digger' of Roman Wroxeter, Thomas Wright, whom John had visited in early May.[43] He heard Prestwich speak at the Royal Society and wrote to *The Athenaeum* on 11 June after reading its report of Evans's lecture. What bothered Wright was the number of stone implements that had been found: 'If we receive them as made by the hands of man, we must suppose that at this extremely remote period the surface of the globe was covered with human beings, who spent all their lives in chipping flints into the rude form of weapons, and throwing them about'.[44] As far as Wright was concerned, the shape of these 'implements' resulted from pebbles being smashed together by the force of the river. Realizing he had no evidence to back up his claim, he nonetheless warned against a current lack of knowledge about the agencies involved to assume 'the absence of agency'.[45]

Evans replied in measured tones a week later.[46] He reported that everyone who had seen the implements from the Somme, such as those who attended Joseph's Flint soirée on 3 May, agreed that 'the majority of them, if not all, were the work of intelligent beings'. He went on, 'It is a hard case if the number of implements that has been found should be turned into an argument against their being implements at all.' Disagreements can be had about how they became incorporated into the deposits of Drift, but not about their significance as human artefacts as opposed to 'some mysterious operation of nature'.

Evans's letter in *The Athenaeum* would have also answered one in the same issue by Richard Cull of the Royal College of Surgeons. He asked if the stone tools had 'been fashioned by art, or are they natural forms?' And then went on, as you might expect, given his institutional background, to wonder at the absence of human remains. No people, no artefacts was his conclusion.[47] On 9 July, however, an old geological friend of Prestwich's, Samuel Pattison (1809–1901), wrote to the magazine *Notes and Queries*, agreeing with Evans rather than either Wright or Cull. He explained the large numbers as a factory, neatly describing its contents as 'the cutlery of the early stone period'.[48]

The reports from the learned societies and the critical letters were read by Samuel Lucas, editor of the brand-new, richly illustrated weekly, *Once a Week*.[49] The first issue came out on 2 July. It opened with a poem by Shirley Brooks that trilled the editorial interests of *Once a Week* for its target female readership. The Amiens discoveries took pride of place on page 3 under the heading *Man among the mammoths*. The piece, with the byline *Alpha*, had reporting inaccuracies (Hoxne became Oxney and Saint-Acheul was located in Abbeville), but Lucas did a good job of recounting the facts for his middlebrow readership. 'Absolutely proven' was his conclusion on the question of great human antiquity. He drew a parallel—shades here of Macaulay—with Pacific islanders, as described by Captain Cook, who lacked metals just like the people who made the tools at Saint-Acheul. The short article ended with a fine engraving of a stone axe from the Drift of Saint-Acheul found by the geologist and cartographer Robert Mylne (1816–90) when he visited with Prestwich and others in late May.[50] It was the first Pleistocene axe to be illustrated in Britain since Frere's letter about Hoxne sixty years earlier. A year later Evans included it in his *Archaeologia* paper.[51]

Then there was a lull. The Rev. Charles Kingsley (1819–75) wrote to Prestwich on 26 August to congratulate him on the Amiens discoveries. They were long expected, he said, but warned him to expect clerical opposition: 'Religious persons will be angry, and try to crush the truth.'[52] But no one showed their hand. Joseph's approach impressed him: 'confining yourself to facts, and building no theories on them'.

On 9 September a short letter appeared in *The Times* under the initials K.P.D.E, a fellow of the Society of Antiquaries. So important were the discoveries, he wrote, that *The Times* was the natural medium to make an inquiry that relates to the whole surface of the globe. K.P.D.E. had three questions for Prestwich and Evans, two of which touched on the artefacts, and one on their age.

His first two questions were:

- Have flint implements been found in similar localities to those already reported?
- Have other implements, not made in flint, been found in geological formations usually regarded as of earlier date than the human race?

Evans replied on 29 September from Nash Mills under the strapline *Man among the mammoths*.[53] Questions one and two were dealt with by mentioning Hoxne, Kent's Cavern,[54] and Brixham, as well as the 'people's' flint from Gray's Inn. In answer to the second, he pointed out that bone tools had been found with ice age animals by Dr Schmerling in the Engis caves near Liège. He finished with a plea for some nineteenth-century citizen science by asking *Times* readers to look for comparable discoveries to the flint illustrated in *Once a Week* and write to him about them. This was the start of his 1872 book *Ancient Stone Implements*.

The grandest reception of the evidence for great human antiquity came on 15 September, when Charles Lyell addressed, as its president, the Geological Section of the British Association for the Advancement of Science. That year this prestigious meeting attended by thousands was held in Marshall College, Aberdeen, in the presence of Prince Albert. Lyell was asked to delay his opening speech so that the prince could attend. The slot was given over instead to a worthy paper on the 'Geology of Aberdeenshire', during which people started to fill up the hall, until by noon and the arrival of the Prince Consort, there were 800 ready to hear the presidential address.[55]

Britain's leading geological authority announced to the prince and the world that great human antiquity was now a fact. Without naming Wright, he answered his objection about the amount of stone artefacts as follows:

The great number of the fossil instruments which have been likened to hatchets, spear heads, and wedges is truly wonderful. More than a thousand of them have already been met with in the last ten years, in the valley of the Somme, in an area 15 miles in length. I infer that a tribe of savages, to whom the use of iron was unknown, made a long sojourn in this region.[56]

Darwin wrote to him five days later: 'I have read with extreme interest in the Aberdeen Paper about the Flint-tools; you have made the whole case far clearer to me: I suppose that you did not think evidence sufficient about

Glacial period.'[57] Lyell did not mention ice in his address. He considered the gravels to be cold-climate and fluviatile in origin, which they are, but in 1859 he linked this, like Prestwich, to a single glacial period first demonstrated in the Swiss Alps twenty years before.[58]

William Pengelly was at the Aberdeen meeting and saw first-hand the reaction to the claims made by what he called drily 'the stone-breaking fraternity'.[59] He was there for the Geological Section meetings that followed Lyell's address a day later and with a smaller audience. There were critical comments, particularly from an ardent supporter of the short chronology, the Rev John Anderson (1796–1864),[60] who spoke 'On Human Remains in Superficial Drift' and lived up to Kingsley's prediction to Joseph to expect religious opposition. He waded through a great amount of rubbish according to Pengelly, a shrewd observer, but nonetheless received 'a very undue share of support from the audience' of geologists.[61] To Pengelly's glee, Lyell replied, handling him as a gentleman before passing him over to Professor John Phillips (1800–74), president of the Geological Society, who 'having rubbed his hands in oil, smoothed him down, but in such a way as to scarify him'. Others corrected his geological misunderstandings about caves, including Pengelly himself, who 'seized him by the collar, dragged him into Brixham Cave, and showed him its facts and their whereabouts'.[62]

November and December saw more correspondence. It began with a letter to the The Times on 18 November from the ever-eager John Wickham Flower under the heading Works of Art in the Drift. He detailed at some length his own discovery at Saint-Acheul.[63] He fully supported Prestwich, as might be expected, adding an entrepreneurial detail: 'Upon our arrival at the pit near Amiens we were met by some little barefoot boys, one of whom, accosted me with the politeness peculiar to his nation, "Monsieur, voulez-vous des langues de chats?"' On replying, 'Oui', Flower was presented with ten or twelve flint axes and money changed hands. With a nod to Frere's 1797 letter, Flower concluded that the axes were so uncouth because they were made by people who had no knowledge of metals.

The next letter appeared in The Athenaeum on 19 November under what at first seems an odd heading, Celts in the Flint, until it is remembered that a celt (pronounced 'selt') was another word for a polished adze and was also used to describe other hafted tools such as axes.[64] The letter came from Professor John Stevens Henslow (1796–1861), Darwin's mentor and friend, who lived at Hitcham in Suffolk and had been to see the pits at Hoxne. He disagreed with Prestwich because the two men he spoke with had a different story to tell. Henslow copied it down because it related to a key part of the evidence, the association of the stone axes with extinct animals: 'They must be very simple folk to think so,' the younger of the two told him, before putting forward his own explanation, which was that a recent factory for flints[65] had become mixed into the sediments. On this testimony Henslow regarded the Hoxne artefacts as humanly made but of no great age. The angular flints which accompanied the axes were either natural fractures or part of an industrial process.[66]

This brought a lengthy reply from Prestwich on 3 December.[67] He recounted in detail what the same men had said to him. A week later he wrote again to clarify the record. He now realized they had spoken to two different old men, whose recollections of the Hoxne pit from fifty years before differed in important details, notably over depths and the association of the axes with different deposits. Prestwich's purpose was not to challenge Henslow so much as to prove that Frere's observations published in 1800 were accurate.[68] The old man Prestwich had spoken with had started work in the pit two years later in 1802, five years before Frere's death.

Henslow had raised doubts but the next, very well-travelled correspondent to *The Times*, cast scorn on the discoveries in a letter published on 1 December.[69] Using the pseudonym SENEX ('old man') he explained the flint implements as the result of rolling together, suddenly and forcibly, a vast amount of rocks, earth, or detritus. This produced heat under which fracturing took place, especially when the rocks were cooling.[70] But what worried SENEX most was the disturbing conclusion if Evans and Prestwich were correct. The stone implements would be a record of an ancient people who in their wanderings had degenerated—lost the arts of manufacture and civilization—from the region 'where man was created'. In this instance, and against all the laws of nature as expounded by Spencer, complex became simple. The moral for SENEX was clear: better to stick with Scripture when it came to assessing human antiquity. Kingsley's warning of religious opposition had another voice.

Two days later it was Evans who responded.[71] He identified two camps who believed either:

- The geological beds are old but the artefacts are not artefacts. This was SENEX's view.

or

- The artefacts are human artefacts but do not belong with the geological beds. Henslow would fit into this camp.

Geologists recognized human handiwork in the examples he and Joseph had brought back from the Somme and rediscovered at Hoxne, while antiquaries did not. The reason for this was simple: the antiquary was used to handling polished axes, and the cruder forms did not fit their idea of human artifice. Following Prestwich, he used Frere's letter and weapons and the Gray's Inn implement as evidence from a pre-scientific age long before the present controversy. These for Evans were objective facts 'unfettered by theory'. Here was a practical Smilean philosophy summed up in the last line of his letter to *The Times*: if you don't believe, then 'Go and see'. Otherwise, 'some ill-meaning persons may employ them as an argument against religion'. SENEX was in his sights. 'But let them rest assured', Evans continued bombastically, 'that truth, whatever it may be, cannot be antagonistic to the cause of religion, while nothing is so prejudicial to that cause as its advocates wilfully and obstinately closing their eyes to the truth.'

Fighting stuff, which drew a second letter from SENEX two days later.[72] Evans was accused of 'clever casuistism [casuistry] and a recourse to the old expedient of "attacking (not to say abusing) your opponent" when advocating a weak cause'. Again, natural causes were invoked by SENEX to explain the shape of the tools.

This time neither Evans nor Prestwich replied. Instead two further letters to *The Times* by the eccentric inventor Robert Collyer, MD (1814-1891), on 9 December and James Wyatt (1816–78) three days later took issue with SENEX. Collyer employed an argument worthy of Smiles to demonstrate human intelligence behind the shape of the stone implements. He began by ridiculing the idea that the heat from volcanos 'chipped' these axes into such regular forms. He was struck instead by the economy of effort in their making. Only those parts needed to make an implement were knapped, and as a result 'all unnecessary labour was purposefully avoided'.[73] And in answer to another of SENEX's questions—where is the evidence of other non-stone artefacts?—Collyer rightly pointed out that they did not survive because they were perishable.

Wyatt's letter repeated Evans's points about antiquarians and geologists before telling *The Times* readership that some years ago he had looked at stone implements in a museum in France and declared 'they were made at a very early period'. The convert Wyatt's letter is the first to mention Evans's new skill of experimental flint knapping.[74]

John had not been idle. Since May he had taught himself to be an expert flint knapper, and this skill brought with it the nickname 'Flint Evans'.[75] Always a quick learner, he celebrated his birthday on 17 November by

making a 'pretty good flint axe'.[76] The skill proved invaluable for confront-
ing flint forgeries and added to his reputation as *the* authority on ancient
stone implements because *he* could make them. On 19 January 1860 he was
on his feet again at the Society of Antiquaries reporting on 'numerous rude
flint flakes' sent to him by John Shelley of Redhill and Reigate.[77] There were
no large axes and nothing polished. There were no extinct animal bones.
These were simply ancient flints. His description was very precise, detailed,
and well illustrated to make his point that irrespective of their age—and he
does not refer to the Stone Age—these were the products of human design.
He assigned them to a period before the Belgic invasion of the Iron Age,
'fashioned by one of the more truly ancient British tribes whose hunting
ground is now traversed by the Brighton Railway, and covered with model
cottages and suburban villages'.[78]

At the same meeting William Vaux (1818–85), a numismatist and expert
on the ancient civilizations of the Near East, reported finds of flint weapons
from southern Babylonia in modern Iraq.[79] He illustrated a finely pointed,
chipped stone axe that would have graced Boucher de Perthes's antediluvian
collection in Abbeville.[80] The geographical spread of human antiquity was
fast expanding.

Towards the end of 1859 the letter writers were running out of steam.
Darwin would soon attract a much wider critique. But this did not stop
another pseudonym, F.G.S. (Fellow of the Geological Society and possibly the
geologist Sir Andrew Ramsay (1814–91)), writing a long letter to *The Times*
on 13 December in an attempt to silence SENEX for good.[81] He recalled
how Prestwich, in a moment of drama, placed the flint he had disinterred
from the Saint-Acheul gravels onto the baize table of the Royal Society. It was
'a flint knife . . . as distinctly wrought by human hands as a "Sheffield whittle"
of Chaucer's or of our own time'. Then, with sarcasm dripping from his quill
pen, he continued, 'Your amiable correspondent "SENEX" may see this knife
without the least trouble, and, like me, somewhat unwillingly satisfy himself
that there are more things in this world than were dreamt of in his philosophy.' Ouch! These were sharp stone tools indeed.

Summer Excursions and a Cold Shoulder

The summer of 1859 saw geologists and antiquarians busily visiting gravel
pits. Charles Lyell arrived in Amiens on 26 July and stayed two days.[82]

His visit added the necessary authority when he addressed the British Association in Aberdeen. Afterwards he corresponded with Prestwich about the species of shells he had dug from the deposits, since these differentiated his Pleistocene and Holocene periods.[83]

In the meantime, Henslow went to Hoxne and talked to the men who worked the pit. In 1859 geologists and antiquaries on both sides of the Channel respected the testimony of the English gravel-sifters and the French *terrassiers*.

Another visitor to Amiens that summer was the evangelical writer Isabelle Duncan (1812–78). She was an ardent creationist, and her most influential work appeared anonymously in early January 1860, *Pre-Adamite Man; or, the Story of our Old Planet and its Inhabitants, Told by Scripture and Science*. She visited Saint-Acheul and brought back stone axes and the sandy matrix they were found in. Examination under the microscope confirmed her presupposition that both sand grains and flints showed signs of abrasion. On this evidence, Duncan dismissed them as human handiwork.[84] She was not going to allow stone tools to contradict the vision of a pre-Adamite race she so passionately believed in. And why? Because Duncan never expected to find any evidence for pre-Adamite Man, who "lived amid the bounties of his Maker, content with the shelter of umbrageous foliage or overhanging rocks [caves]. A worshipper rather than a worker, he honoured God by praise and service.'[85] Not being a worker meant the pre-Adamite had no need of stone tools. Her purpose in going to Amiens was not to 'Go and see' as Evans exhorted his readers, but rather to 'Go to not see'. Even so, Duncan and her evangelical followers needed the pre-Adamites to fill the hole in time opened up by geologists. Therefore, and rather confusingly, the idea that Adam was born of pre-Adamite parents had some evangelical support: 'Adam had a navel, for Adam had ancestors.'[86] There was reason to celebrate the birth and not the death of Adam at the hands of geological evidence.[87] He just didn't need stone tools.

It was now the turn of French scientists, spurred by the reports from Prestwich and Evans, to visit the Somme. They arrived in August. The young Georges Pouchet (1833–94), at Lyell's urging, was sent by his museum in Rouen to report on the Amiens stone tools.[88] He toured the Saint-Acheul pits with the workmen, found stone tools in the sediments, just as he expected to, and was shown the two photographs taken for Prestwich and Evans.[89] Convinced by all he saw, he left Amiens with copies of the pictures, thanks to Pinsard, and the obligatory bag of stone implements.[90]

The geologist Albert Gaudry (1827–1908) examined the Amiens evidence from 7 to 9 August.[91] He sent a letter with his findings to the Académie des Sciences in Paris, and on 26 September an extract was published as a brief note in their *Comptes-Rendus*, equivalent to the Royal Society's *Proceedings*.[92] Here Gaudry declared that the association of stone tools and extinct animals was verified. To reach this conclusion, he dug a deep trench in the Fréville pit, never letting the workmen out of his sight to eliminate the possibility of fraud. He was rewarded with nine axes *in situ* in the water-laid deposits and closely associated with remains of an extinct species of horse and aurochs, wild cattle. Gaudry declared, 'The precise determination of the deposit where the axes were found proves definitively that man was a contemporary of many of the great fossil animals now extinct.'[93] A fuller account of the geology of Fréville's pit at Saint-Acheul came later.[94] He had witnesses present—Charles Pinsard and Jacques Garnier, who had been there in April, and the architect and academician Jacques Hittorff (1792–1867).[95] He widened his report to take in other pits at Amiens and the work of the geologist Buteux along the line of the Amiens to Boulogne railway. This allowed him to expand the list of ice age fauna to include rhino, mammoth, and hippopotamus. Gaudry was supported in the Académie on 10 October by Pouchet in a letter read out by the zoologist Isidore Geoffroy-Saint-Hilaire (1805–61).[96]

Gaudry also examined in detail one of Rigollot's claims that some small pierced pebbles found in the Saint-Acheul pits were beads on a necklace. He correctly identified them as tiny fossil sponges that are common in the chalk. He put some of the pebbles under his microscope to look for traces of manufacture around the perforation. None was visible.[97]

Gaudry's brief communications to the Académie had one glaring omission. Not once were Boucher de Perthes and his books mentioned.[98] Contrast this slight with the reports in England by Evans, Prestwich, and Lyell, as well as letters to newspapers where Rigollot and Boucher de Perthes were widely praised for their originality, zeal, and perseverance. Incensed, Jacques wrote a long letter on 11 October to one of his few supporters in the Académie, Geoffroy-Saint-Hilaire. He chronicled the history of his discoveries, stating that 'the first verification of my claims was carried out by Prestwich and Evans in April 1859. The second by Godwin-Austen, Flower and Mylne. The third by Lyell.'[99]

Quite rightly, Jacques asked for the oversight in Gaudry's account to be corrected. He did not think the omission was intentional. It seems likely that Gaudry was not at fault, but rather Jacques's long-time opponent

Jean-Baptiste Élie de Beaumont (1798–1874), who edited the letter sent to the Académie and chose to write him out of history.

With a bit of pushing and shoving Jacques now had his deserved *gloire*. His letter to the Académie was published in the *Comptes-Rendus*, this time edited by Geoffroy-Saint-Hilaire.[100] Immediately below was a brief note by the geologist Élie de Beaumont pointing out that at the meeting on 3 October Boucher de Perthes work had been mentioned but not recorded.[101] The reason, he went on, was because there was nothing new to report; communications by Boucher de Perthes on his antediluvian finds had been published in the *Comptes-Rendus* as far back as 1846. This half apology called for a further response. On 31 October Élie de Beaumont was on his feet again at the Académie des Sciences, this time reading out a letter written to him by Prestwich. In this Joseph detailed the circumstances of the discovery on 27 April, giving Jacques full credit.[102] Prestwich finished with a call to arms following the demonstration that humans lived alongside extinct animals:

> A fact that has put geology and ethnology *en rapport*, for the first time, and which, because of the questions now raised, cannot fail but resolve problems of a very complicated geological nature and so establish new scientific truths.[103]

Not everyone was as keen as Prestwich and Evans on bringing together geology, anthropology, and archaeology. In June, Robert Godwin-Austen, who by then had visited Jacques with Mylne and Flower, wrote to Prestwich warning him that their case could be damaged by 'bad evidence and worse reasoning'. This would be the fate of geologists like them if they allied too closely with the antiquary and the anthropologist. Beware, he wrote to his friend: 'Falconer and Evans are to us what the two cunning Greeks were who conducted the fatal horse into Troy,'[104] At least Boucher de Perthes was no longer their Achilles heel. Instead, turning Paris' bow against the capital, he was triumphantly announcing to Abbeville's Société d'Emulation that all his work, and their faith in him, their president, had been vindicated.[105]

How Old *Was* Old?

The paradox of the time revolution of 1859 is that nobody knew how old it was. The outgoing president of the Geological Society, John Phillips, pronounced in February 1860 that geological time eluded the grasp of the imagination.[106] The amount of time represented was beyond scientific reckoning. He gave an example: 16 feet in a gravel pit in Oxfordshire compressed the eras of historic, prehistoric, and pre-Adamitic time, with mammoths at the bottom and the footprints of Charles I fleeing the parliamentary army at the top.[107] In the year before the time revolution John Kemble wrote of the 'infinite distance from our own' of deposits like the Somme, where 20 feet below the surface rhino bones had been found. They represented an incalculable lapse of centuries.[108]

Age was not an issue, however, for those working within an historic time scale. Isabelle Duncan had no doubts that Adam was created 6,000 years ago and 2,000 years later came Noah's Flood (Table 4.1).[109] She equated the seven biblical days with geological epochs and where divine creation fitted everything together in her ingenious interpretation of Mosaic history. As described by the historian of science Stephen Snobelen, what she achieved was the Christianization of geological time through a strategy of 'temporal imperialism'.[110] This saw her turn geological eras into the first six days of Genesis. Her goal, reconciling Scripture with science, illustrated the truth of the former.[111] SENEX took a similar view, where the rotations of the days in Genesis lasted an 'undefined time'.[112]

Samuel Lucas in *Once a Week* castigated the unworkable biblical time-scale of 6,000 years. Instead, he argued, the discoveries of Jacques, Joseph, and John showed the earth was inhabited at an 'enormously remote period', how old being indicated by the 'indisputable fact that man is older than the present distribution of land and water, hill and valley.'[113] This was also Charles Goodwin's view in the heretical *Essays and Reviews* of 1860.[114] This polymath, Anglo-Saxon scholar and Egyptologist, linguist, lawyer, geologist, and music critic[115] inspected the geological evidence of the Drift at first hand and was impressed by the amount of time it represented. Furthermore, it contained evidence of abundant life so that:

> No blank chaotic gap of death and darkness separated the creation to which man belongs from that of the old extinct elephant, hippopotamus and hyaena.[116]

Elsewhere in the same paper he acknowledged the recent demonstration of the antiquity of humans (Prestwich and Evans) and the development of species (Darwin and Wallace). These advances, as he called them, were based on evidence such as these Pleistocene animals, 'undeniable existing

Table 4.1 The Mosaic chronology

	A.M. (Anno Mundi) (Year after Creation)	AD–BC	
1857	6000	1857 Present day	
Destruction of Jerusalem by the Romans	4213	71 AD	
Crucifixion	4176	34 AD	
Birth of Christ	**4142**	**0 AD**	
Death of Alexander	3821	321 BC	
Release of the Jews by Cyrus	3584	558 BC	
Sack of Jerusalem by the Sassanians	3514	628 BC	
David crowned king	3000	1142 BC	
Fall of Troy	2917	1225 BC	
Exodus from Egypt	2453	1689 BC	
Ninus the Younger	2000	2142 BC	Son of Ninus
Semiramis	1958	2184 BC	Assyrian queen
Abraham born	1948	2194 BC	
Ninus	1906	2236 BC	Founder of Nineveh
Belus	1844	2298 BC	Assyrian king
Nimrod	1788	2354 BC	Great-grandson of Noah
Tower of Babel	1742	2400 BC	Language diversity begins
The Deluge (Noah's Flood)	**1656**	**2486 BC**	Racial diversity begins
Noah born	1056	3086 BC	
Enoch translated	987	3155 BC	aged 365
Lamech born	874	3268 BC	Father of Noah
Methuselah born	687	3455 BC	Grandfather of Noah
Enoch born	622	3520 BC	Great-grandfather of Noah
Jared born	460	3682 BC	Father of Enoch
Mahalaleel born	395	3747 BC	Father of Jared
Cainan born	325	3817 BC	Oldest son of Enos
Enos born	235	3907 BC	Oldest son of Seth
Seth born	130	4012 BC	Third son of Adam and Eve
Adam created	**0**	**4142 BC**	

The biblical chronology of 6,000 years according to Samuel Pattison (1858: 66–7, with explanations of the names added). Like Prestwich, Evans, and Lubbock, he was a fellow of the Geological Society and regarded geological eras as 'waymarks of pre-historic time' (1858: 58). His book *The Earth and the Word: Or Geology for Bible Students* sought to bring science and Scripture together. Building on the work of the Rev. Franke Parker, he used Genesis, as Archbishop James Ussher had done two hundred years previously, to make a chronology by combining the ages of long-lived patriarchs such as Enos and Lamech. However, unlike Isabelle Duncan, who attempted a similar rapprochement in *Pre-Adamite Man* (1860), Pattison concluded that no evidence for humans before Adam had, or would be found (1858: 67). The earth existed before the Creation but human history started with it or, as he put it, 'Palaeontology and mineralogy tell us the world has a history not recorded' (1858: 76). This left plenty of room for geologists like him to study the formation of the world and leave Scripture to account for its greatest creation, humans. The Tower of Babel and Noah's sons who survived the Flood were the two events that explained the variety of nineteenth-century global ethnicity (Trautmann 1992: 386).

monuments', that contradicted the revealed truth of Genesis and the Mosaic account of history.[117]

The issue of how old *was* old was approached in characteristically jaunty fashion by Evans in his *Archaeologia* paper. He introduced his readers gently to the topic with a fragment of verse:

> Antiquity appears to have begun
> Long after their primeval race was run.[118]

The lines came from Horace Smith's poem, *Address to the Mummy in Belzoni's Exhibition*. The allusion was to a history before antiquity, a deep human history, be it either prehistoric or pre-Adamitic. In 1825, when Smith penned his poem, the historic era went back as far as the Egyptians. The lists of kings and pharaohs set down by the Egyptian Manetho in the third century BC extended the historic chronology beyond the 6,000-year-old barrier.[119] Antiquity was reserved for Romulus and Remus and classical Rome. In the next stanza Smith asked the Mummy:

> Thou couldst develop—if that withered tongue
> Might tell us what those sightless orbs have seen—
> How the world looked when it was fresh and young,
> And the great Deluge still had left it green:
> Or was it then so old that history's pages
> Contained no record of its early ages?

Capturing the spirit of Smith's poem, Evans presented his discovery of flint implements wrought by the hand of man in undisturbed deposits of sand and gravels as 'furnishing the earliest relics of the human race with which we can hope to become acquainted'.[120] Relics, he implied, from so remote a period that Ancient Rome and the even older Pharaohs seem young. And although he doesn't quote this stanza, he would have known that what he and Joseph, following the lead of Boucher de Perthes, had done was to find the first record of early ages on history's pages.

Adjectives, however, were all that the geologists and antiquaries had as they pushed back the boundaries of prehistoric and pre-Adamitic time. 'A vast lapse of ages separating the era in which the fossil implements were framed and that of the invasion of Gaul by the Romans'[121] was how Lyell described the discoveries of stone tools to the large audience at the British Association.

K.P.D.E.'s third question in his letter to *The Times* had tried to put Evans on the spot—had any human remains ever been found 'in a position indicative of a greater antiquity than that usually conceded?' By which he meant older than 6,000 years. Evans's reply was full of bluster. Anyone like himself who had studied the evidence must concede that it established beyond doubt 'a greater antiquity of the human race than is consistent with our ordinary chronology; though to what extent its bounds are to be transcended must remain an open question'.[122]

Wickham Flower was similarly puzzled when confronted with an antiquity greater than anything previously suspected. He felt at a loss to imagine the human contemporaries of the mammoth and hippopotamus, 'as Robinson Crusoe was perplexed by seeing the footprints of his mysterious visitor in the sands of his desert island'.[123]

How old did Evans and Prestwich believe their flints to be? The short answer is they had no idea and, perhaps surprisingly, were not that bothered. Boucher de Perthes had floated the figure of 'thousands of centuries'[124] for his antediluvian stone tools and the extinct beasts which accompanied them. Joseph and John both wrote that speculation on this matter would be irresponsible and regarded Jacques's figures as outrageous. But without some estimate, how could they claim, as John did, that their discoveries will engage the 'attention of all who, whether on ethnological, philological, or theological grounds, are interested in the great question of the antiquity of man upon the earth'.[125]

Joseph wound himself into the question in the closing paragraphs of his *Transactions* paper. He reiterated that Boucher de Perthes and Rigollot were correct to argue for a great antiquity on the basis of these worked flints, but he then dismissed their diluvial theory of a global catastrophe that wiped out the animals and humans.[126] After the meticulous geological detail and careful presentation, all he had to fall back on was familiar adjectives. The extinction of the great ice age mammals—mammoth and rhinoceros—required 'a great lapse of time', and this took place during the 'latest of our geological changes at a remote, and, to us, unknown distance'. In the working manuscript of his *Transactions* paper a 'vastly' once qualified 'remote', but this was crossed out and did not make it into print. It was a case of cold feet. Joseph had looked down the long corridor of deep human history and found it unsettling. He continued: the present evidence 'does not seem to me to necessitate the carrying of man back in past time, so much as the bringing forward of the extinct animals towards our own time'.[127] Just at the moment when the telescope of human history was extending, he snapped it shut.

Evans agreed with his friend, using exactly the same argument in *Archaeologia*.[128] However, while Joseph favoured caution and moved everything towards the present, Evans was wary of coming to a 'hasty decision'.[129] Having gone over the same ground several times in rather ponderous fashion he reiterated his unshakeable view that in a period of antiquity 'remote beyond any of which we have hitherto found traces this portion of the globe was peopled by man'.[130] And not only that, but these humans witnessed the geological changes of ice and water that led to the formation of the beds of Drift.

A few scientists did hazard estimates of how old *was* old. One was made for Egypt by the educational reformer and geologist Leonard Horner (1785–1864), who succeeded Phillips as president of the Geological Society.[131] In a paper in the *Transactions* in 1858 Horner calculated the annual rate at which the Nile deposited its silts to work out the age of a fragment of pottery that he had found at the bottom of a trench 39 feet deep.[132] The depth indicated to Horner an age of 11,517 years before the Christian era, over 13,000 years before 1854, when he was doing his research. This marked the appearance of people in this part of the ancient world. Horner cited in support the German scholar, Chevalier Bunsen (1791–1860), whose documentary researches led him to conclude that a long period preceded the Ancient Egyptians and that the history of the human race began 20,000 years ago. Manetho was one of his sources. Darwin was an enthusiastic referee for Horner's paper.[133] A year later, however, a Roman tile was found beneath the pottery shard, and the calculation had to be abandoned. The idea of using rates of sedimentation by rivers as a way of calculating time did, however, continue.[134]

In both of his letters to *The Times*, SENEX ridiculed Horner and Darwin for their erroneous ages.[135] He saw no evidence that changed the 6,000-year boundary to human history. The six days of rotation in the account in Genesis were the geological periods that prepared 'this world for the abode of man'. He also disparaged Manetho's chronology, who, he claimed, recorded a flood that undermined his case for such extravagant ages. Furthermore, SENEX asserted, in 1859 more of the world was uninhabited than settled. Why would this be the case, he asked disingenuously, if a chronology of 14,000 years was in any way accurate? More than enough time to fill up the globe. In his view, the lack of humans in many regions proved the short biblical chronology of 6,000 years.

This was all too much for Darwin. He knew who SENEX was: Vice-Admiral Robert Fitz-Roy, his Captain on the Beagle and now in charge of the nascent shipping forecast at the meteorological office.[136] He wrote to Charles Lyell on 15 December: 'it is a pity he [Fitz-Roy] did not add his theory of the extinction of Mastodon, etc., from the door of [the] Ark being made too small. What a mixture of conceit and Folly, & the greatest newspaper in the world, inserts it!'[137]

In his reply to SENEX, Collyer expressed his bafflement by the issue of age: 'whether 10,000 or 100,000 years ago, I cannot understand how the question of their reality as "works of art" is affected thereby'. Age had nothing to do with accepting artefacts as made by humans. Elsewhere, the same argument could have been applied to Phillips's unimaginable span of geological time. The Weald of Sussex and Kent had been eroded in 1,332,000 years according to Phillips in a calculation that could have been written on the back of one of Evans's envelopes. Darwin had applied a much slower rate of erosion and reached an estimate of 306,662,400 years, an inconceivable number in Phillips's view.[138] But the length of time taken was irrelevant. The Weald *was* eroded and could be studied as a geological process with geological methods without knowing how long it took.

This was Charles Goodwin's view in his paper in *Essays and Reviews:* 'No well-instructed person', he wrote, 'now doubts the great antiquity of the earth any more than its motion.'[139] He bemoaned an educational system that taught England's children that the Earth went round the Sun but was made in six days and less than 6,000 years old. Instead geology showed that those years had to be counted in millions, not thousands, from the first creation to the appearance of humans.

Others found SENEX extremely irritating. F.G.S. did not like the way SENEX sneered at those who used Manetho and pulled him up for claiming that the Egyptian mentioned a flood, 'which he certainly does not'.[140] Instead, and citing Lyell, F.G.S. agreed that the evidence of the Somme showed that the people who made the stone implements occupied the region before the Straits of Dover were cut. History began not just before Adam in a world of ice and woolly mammoths. It lasted also for a very long time before Britain became an island. And yes, those stone tools were made and left by ancient people known only by their handiwork and not their bones.

By the end of the year Prestwich and Evans's case for human antiquity had been thoroughly reviewed and a ringing endorsement sent to *The Athenaeum* on 31 December by the leading Danish antiquary Jens Worsaae. He wrote at length, concluding there were no grounds to doubt the tools as the handiwork of very ancient humans. He went on, 'I feel convinced we are at the commencement of some of the most remarkable discoveries which have been lately made.'[141] Scientists and antiquaries would be emancipated from 'old and new prejudices, and from so-called historical theories'. The prehistory of the Northern Antiquaries was vindicated by the time revolutionaries of the Somme.

Despite some resistance, the people whose opinion they valued were agreed: 6,000 years was no longer enough for the beginning of human history; Adam had ancestors and stone implements stood as their proxies. Simple tools marked the start of human history. The first phase of the time revolution was complete. It was now time to consolidate.

Notes

1. Cannadine 2017: Kindle p. 4641. The 1851 census revealed that out of a population of almost 18 million, only 5.2 million attended Church of England services, while 4.9 million attended other Christian places of worship. The latter were predominantly Evangelical and included Methodists, Baptists, and Quakers; see https://www.english-heritage.org.uk/learn/story-of-england/victorian/religion/, accessed 31 July 2020.
2. As Thomas Carlyle put it, the sentry who deserts his post should be shot (Altholz 1982: 188).
3. Rowland Williams and Henry Bristow Wilson were acquitted by the Privy Council in 1864 of the three counts of heresy against them, but not before they had been suspended from their duties (Beckwith 1994: 2).
4. Ellegard 1958: 27 and 106.
5. The legacy continues to appal. The theologian Roger Beckwith (1994: 9) writes that in its unbelieving approach to the Bible, *Essays and Reviews* 'has hindered the mission of the church ever since. It lies at the root of many of the calamities which have afflicted the church in our own day, and from which, until we repent of unbelief, the church will never recover.'
6. Pizanias (2011) and Defrance-Jublot (2011) describe the Protestant and Catholic reactions in France to the time revolution of 1859.
7. Temple 1860.
8. Spencer, Letter to Mill, 17 February 1859 (Duncan 1911: 93). Spencer argued that liberty must not be regarded as synonymous with democratic government. Just the reverse. He accused democracy of increasing the tyranny of the state over individuals.
9. In the 1859 election 565,000 votes were cast, about 3 per cent of the total population.
10. Buckle 1859: 538.

11. Buckle 1858a: 15. His unfinished *History of England* sought to understand the causes of progress based on the dynamic formula that nature influenced human actions as much as humans acted on nature. Cannadine (2017: Kindle p. 5822) regards Buckle's history as one of the key works for understanding the mood and mind of the mid-Victorian generation.

12. Wilson 2016.

13. Spencer's essay *Progress: Its Law and Cause* was first published in the *Westminster Gazette* in April 1857 and was reprinted in his *Essays* a year later (1858: 3). Spencer uses the term evolution here in 1857, while two years later in *The Origin*, Darwin does not. For Spencer his law of progress from simple to complex can be seen 'from the earliest traceable cosmical changes down to the latest results of civilisation' (1858: 3). His view of acquired characteristics owes far more to Lamarck than to Darwin (Bowler 2009).

14. Darwin and Wallace 1858.

15. Buckle (1859: 538) presented his historical proof for progress as follows:

> as civilization advances, the impunity and rewards of wickedness diminish. In a barbarous state of society, virtue is invariably trampled upon...In that stage of affairs, the worst criminals are the most prosperous men. But in every succeeding step of the great progress, injustice becomes more hazardous...A large part of the power, the honour, and the fame formerly possessed by evil men is transferred to good men.

16. Patton (2016: Kindle p. 224) describes Lubbock's heartfelt adherence to Anglicanism as vaguely defined.

17. Macaulay 1848–55: 3.

18. Macaulay 1848–55: 2.

19. Margaret died in 1893 at the age of 45. Arthur Evans never remarried.

20. Freeman 1867: 1.

21. John Evans did not care for Freeman's great book (Joan Evans 1964: 16) probably because its many facts were marshalled to prove a theory.

22. Bremner and Conlin (2015: 9) characterize Freeman's dramatic style as 'histrionics, not history'.

23. Sellar and Yeatman 1931.

24. Chippindale 1988; Rowley-Conwy 2006.

25. Worsaae 1849: 149–50.

26. Kemble 1857: 478.

27. Rowley-Conwy 2006: Table 1.

28. Rowley-Conwy 2007: Fig. 3.5. In 1838 the Swedish archaeologist Sven Nilsson turned the three-age model into economic stages. The Stone Age became hunters and the Bronze and Iron Ages pastoralists and farmers.

29. The anthropologist and philosopher Ernest Gellner (1986: 7) named this, tongue in cheek, the doctrine of Trinitarianism. For the antecedents of the three-age system in the Enlightenment, see Meek (1973, 1976).

30. Morse 1999: 1. The book is found in many UK libraries but catalogued under the name of its translator (Reviewer 2).

31. Kemble 1857: 467. In 2003 the Saudi Arabian minister to OPEC, Sheikh Yamani, made a similar observation: 'The Stone Age did not end for lack of stone, and the Oil Age will end long before the world runs out of oil' (*The Economist*, 23 October 2003).

32. Morgan employed a three-age format, from Savagery (hunting) to Barbarism (herding and farming) to Civilization (towns, cities, and commerce), that depended on the changes from stone to bronze to iron. This scheme was not original but has been much copied since. Morgan is known as Marx's anthropologist, and his originality lay in his knowledge of American ethnography; see https://www.marxists.org/reference/archive/morgan-lewis/ancient-society/ch01. htm, accessed 31 July 2020. Trautmann (1992: 381) points out that the time revolution of the 1860s changed the way Morgan conceptualized his scheme for social evolution.

33. Evans in Prestwich 1860: 312.

34. Evans 1860a: 293.

35. 'Celtic', if it means anything chronologically, would refer to the Iron Age and not what Lubbock in 1865 would call the Neolithic (New Stone Age), with its distinctive polished 'celts' (pronounced 'selts'), used either as chisels or axes (see Chapter 6).

36. Evans 1860a: 293.

37. Panizzi 1859. The antiquities section was written by a senior curator, Edward Hawkins.

38. The Natural History Museum in South Kensington was founded in 1881 with these collections.

39. Caygill and Cherry 1997.

40. On loan from the Society of Antiquaries.

41. Society of Antiquaries, 2 June, *The Athenaeum*, 11 June 1859, pp. 781–2.

42. *The Gentleman's Magazine*, July 1859, vol. 207: 45–8.

43. A glance at Wright's list of publications in Wikipedia suggests he is England's answer to Boucher de Perthes. Books on Latin poems, Piers Ploughman, Queen Elizabeth, narratives of sorcery and magic—the list is polymathic and poorly focused. Wikipedia lists thirty-seven books, none of them on his excavations at Wroxeter, although he did write a short guidebook to the site in 1872.

44. T. Wright, Letter to *The Athenaeum*, written on 11 June, published on 18 June 1859, p. 809.

45. *Darwin Correspondence*, Darwin to Hooker, 22 June 1859. He endorsed Wright's comments about the numbers and reflected that Boucher de Perthes's poor

drawings led him to conclude they were angular stone fragments broken by ice action.

46. Evans, Letter to *The Athenaeum*, written on 21 June, published 25 June 1859, p. 841.

47. Richard Cull, Letter to *The Athenaeum*, 25 June 1859, p. 841. His point was answered by Lubbock.

48. *Celtic Remains in Jamaica*. S. R. Pattison, Letter to *Notes and Queries: Medium of Inter-Communication for Literary Men*, 9 July 1859, p. 207. Samuel Rowles Pattison was an old friend of Prestwich (G. A. Prestwich 1899: 39) and a fellow of the Geological Society since 1839. In 1858 he published *The Earth and the Word; Or, Geology for Bible Students*. His reaction was typical: belief in geological ages for extinct faunas but belief in the 6,000-year-long chronology for humans. Prestwich, deep down, shared these sympathies.

49. *Once a Week*, 1859, pp. 3–4. Human antiquity was the splash which opened this lavishly illustrated weekly owned by William Bradbury and Frederick Evans, who also owned *Punch*. In 1859 Evans acrimoniously parted company with Dickens, who branched out with *All the Year Round*, where he serialized *A Tale of Two Cities*. *Once a Week* was edited by Samuel Lucas and directed at a middlebrow, middle-class readership.

50. The full party was Mylne, whom Prestwich knew through his geological maps of London, Godwin-Austen, and Flower. This took place after Joseph's paper to the Royal Society on 26 May.

51. Evans 1860a.

52. G. A. Prestwich 1899: 136.

53. Evans also took the opportunity to dismiss another letter to *The Times* by Mr Tindal of Bridlington, who drew attention to similar artefacts in Yorkshire. They needed more investigation.

54. A whimsical review of the report on Kent's Cavern by the late Rev. MacEnery, edited by E. Vivian from his notes, appeared in *The Athenaeum* on 30 April 1859. There is mention of primeval savagery but nothing about antediluvian humans.

55. Lyell, Letter to his wife, 15 September 1859 (Lyell 1881: 323–4).

56. Lyell 1860: 94.

57. *Darwin Correspondence*, Darwin to Lyell, 20 September 1859.

58. Louis Agassiz (1807–73) first proposed a widespread glacial period in 1837 (Imbrie and Imbrie 1979).

59. Pengelly, Letter to his wife, 19 September 1859 (Pengelly 1897: 91).

60. Anderson was an amateur geologist and a fellow of the Geological Society of London, known for his work on fossils from the yellow sandstone. His belief in the Mosaic timescale and divine creation for humans can be found in his book, *The Course of Creation* (1854).

61. Pengelly, Letter to his wife, 17 September 1859 (Pengelly 1897: 90).

62. Pengelly, Letter to his wife, 17 September 1859 (Pengelly 1897: 90).

63. Flower, Letter to *The Times*, 18 November 1859, p. 7. The letter was sent nine days earlier.

64. https://en.wikisource.org/wiki/1911_Encyclop%C3%A6dia_Britannica/Celt_(tool), accessed 31 July 2020.

65. Presumably gun flints, but Henslow does not say.

66. *The Athenaeum*, 19 November 1859, p. 668.

67. Written on 1 December, published *The Athenaeum*, 3 December 1859, pp. 740–1.

68. *The Athenaeum*, 3 December 1859, 'Flint Implements in the Drift', pp. 740–1; 10 December 1859, pp. 775–6.

69. *The Times*, 1 December 1859, 'Works of Art in the Drift', p. 8.

70. Darwin referred to this as Fitz-Roy's pebble theory, about which he may have heard more than he wanted to when they were aboard *The Beagle* in the 1830s. *Darwin Correspondence*, Darwin to Lyell, 3 December 1859.

71. Evans, 'Works of Art in the Drift', Letter to *The Times*, 3 December 1859, p. 10.

72. SENEX, 'Broken Flint in "Drift"', Letter to *The Times*, 5 December 1859, p. 11.

73. 'Flint implements in the Drift', Letter to *The Times*, 9 December 1859, p. 7.

74. 'Flint implements in the Drift', Letter to *The Times*, 12 December 1859, p. 10. Wyatt answered Evans's call for people to search their local gravel pits for Pleistocene axes, reporting many from the Ouse valley in Bedfordshire in the 1860s.

75. Evans 1943: 107. Lamdin-Whymark (2009) explores Evans's experimental flint knapping.

76. Evans 1943: 105, Fanny Phelps diary entry.

77. *Proceedings of the Society of Antiquaries* 1859-61: 69–77. These stone tools are now classified as Mesolithic.

78. *Proceedings of the Society of Antiquaries* 1860: 77.

79. *Proceedings of the Society of Antiquaries* 1860: 64–9.

80. *Proceedings of the Society of Antiquaries* 1860: 66.

81. F.G.S., 'Works of Art in the Drift', Letter to *The Times*, 13 December 1859, p. 5. In a letter to *The Athenaeum* on 16 July 1859, the geologist Professor Andrew Ramsay of UCL, and later president of the Geological Society, stated that the Amiens and Abbeville stone hatchets were 'as clearly works of art as any Sheffield whittle', which is the only clue as to the identity of F.G.S. (see Lubbock 1862: 247).

82. Hurel and Coye 2011.

83. Prestwich, Letter to Lyell, 11 August 1859 (G. A. Prestwich 1899: 134–5).

84. Duncan, Preface to the third edition of *Pre-Adamite Man*, March 1860.

85. Duncan 1860: 138–9.

86. Livingstone 2008: 137.

87. Livingstone 2008: 137.

88. Remy-Watté 2011: 226.
89. De Bussac 1999: 300; Pouchet 1860.
90. Remy-Watté 2011: 226–9. Some of the twenty artefacts he brought back to Rouen are shown in Remy-Watté (2011: Figs 8 and 9).
91. Hurel and Coye 2011.
92. Gaudry 1859a: 453–4.
93. Gaudry 1859a: 454, 'La détermination precise du gisement des haches prouve définitivement que l'homme a été contemporain de plusieurs des grand animaux fossils détruits de nos jours.'
94. Gaudry 1859b: 465–7.
95. Hurel and Coye 2011: 18.
96. Pouchet 1859. He visited Saint-Acheul after Gaudry on 25 August.
97. Evans in his reply to K.P.D.E.'s second question about non-flint artefacts came to the same conclusion about these 'stone beads'. Letter to *The Times*, 29 September 1859.
98. Gaudry 1859a, 1859b.
99. Letter translated by Paul and Anny Mellars (Daniel 1972: 317–20). See also Agache (1974).
100. *Comptes-Rendus* 1859, 49: 581. Geoffroy-Saint-Hilaire had presented Jacques's findings to an unappreciative Académie in May 1858.
101. *Comptes-Rendus* 1859, 49: 581–2.
102. *Comptes-Rendus* 634–6.
103. *Comptes-Rendus* 636: 'un fait qui vient de mettre pour la première fois la géologie et l'ethnologie en rapport et ne peut pas manquer, par les questions auxquelles ça donnera lieu, d'arriver à résoudre des problèmes en géologie très-compliqués et à établir des nouvelles vérités dans la science'.
104. Godwin-Austen, Letter to Prestwich, 13 June 1859 (G. A. Prestwich 1899: 128).
105. Boucher de Perthes 1864b. The meeting of the Abbeville society took place on 7 June 1860.
106. Phillips 1860: li. One of the fellows of the Geological Society, Samuel Pattison (1858: 68), expressed a similar opinion about geological time and 'our inadequacy to conceive of the immense duration required'.
107. Phillips 1860: lv.
108. Kemble 1857: 464.
109. Snobelen 2001: 84; Duncan 1860: 46 and 119.
110. Snobelen 2001: 97.
111. All of this was presented in a lively illustration showing four distinct strata in deep time. Adam's race, indicated by a pyramid, is at the top. Then came a layer of lifeless ice beneath which in a pre-Adamitic world mammoths roamed. Finally dinosaurs are shown in the lowest stratum. Mosaic history was an immensely malleable concept. It allowed Duncan to acknowledge the possibility of a pre-Adamite race of humans, without being convinced by the artefacts

from the Somme, and in this way the deep past was drawn into a quasi-biblical timeframe (Livingstone 2008: 89).

112. SENEX 'Broken Flint in "Drift"', Letter to *The Times*, 5 December 1859, p. 11.

113. *Once a Week*, 1859, vol. 1, p. 3.

114. Goodwin 1860: 207–53. *Essays and Reviews* was seen by its varied authors as 'an attempt to illustrate the advantage derivable to the cause of religious and moral truth, from a free handling, in a becoming spirit, of subjects peculiarly liable to suffer by the repetition of conventional language, and from traditional methods of treatment'. Progressive in tone, it dealt with weighty Christian issues but did not directly address the antiquity of humans. Goodwin argued it was time to move on from Mosaic accounts of creation before the appearance of humans. He concluded that 'no one contends that it [the Mosaic timetable of days of creation] can be used as a basis of astronomical or geological teaching' (p. 253). Genesis now had to be treated as a historical document not an account of history.

115. *Dictionary of National Biography*.

116. Goodwin 1860: 229.

117. Goodwin 1860: 229. There were several rejoinders to his paper, but Goodwin never replied.

118. Evans 1860a: 281. The poem by Horace Smith (1779–1849), *Address to the Mummy in Belzoni's Exhibition*, was first published in 1825. It became a staple, reprinted many times over the years in papers such as *The Penny Magazine of the Society for the Diffusion of Useful Knowledge* and the *New Monthly Magazine*. In it Smith interrogates the Egyptian mummy that Belzoni had on display in London in 1821. Smith wants to know answers to questions such as who built the pyramids and 'Is Pompey's pillar really a misnomer?' 'Perhaps thou wert a Mason, and forbidden / by oath to tell the secrets of thy trade.' The poem was also used as a children's song, and is Smith's best-known work. In his *Archaeologia* paper Evans misquotes, substituting 'their' for 'thy' in 'thy primeval race'.

119. For a description of the little we know about Manetho, his *Aegyptiaca*, and its importance for Egyptologists, see https://en.wikipedia.org/wiki/Manetho, accessed 31 July 2020. His ages for the gods, demigods and Egyptian spirits of the dead alone totalled 24,925 years. The decipherment of Ancient Egyptian hieroglyphics following the discovery of the Rosetta Stone in 1799 was the first assault on the 6,000-year-long Mosaic chronology and in retrospect paved the way for the time revolution (Trautmann 1992: 387).

120. Evans 1860a: 281.

121. Lyell 1860: 95.

122. Evans, 'Works of Art in the Drift', Letter to *The Times*, 3 December 1859, p. 10.

123. Flower, 'Works of Art in the Drift', Letter to *The Times*, 18 November 1859, p. 7.

124. Boucher de Perthes 1847: 173.
125. Evans 1860a: 281.
126. Prestwich 1860: 308.
127. Prestwich 1860: 309.
128. Evans 1860a: 303.
129. The period 8,000 to 10,000 years ago was the age in 1864 that Prestwich later settled on for the age of the extinct animals (Chapter 5).
130. Evans 1860a: 306.
131. Horner became Charles Lyell's father-in-law when he married his daughter Mary in 1832.
132. Horner 1858.
133. *Darwin Correspondence*, Darwin to the Secretary of the Royal Society, 22 March 1858. By recommending Horner's paper for publication, Darwin, in Fitz-Roy's eyes, agreed with this age for human ancestry.
134. See Chapter 5.
135. SENEX also added Bunsen to his list of wrong-headed timekeepers.
136. See Chapter 1.
137. *Darwin Correspondence*, Darwin to Lyell, 3 December 1859.
138. Phillips 1860: li–lii.
139. Goodwin 1860: 210.
140. F.G.S., Letter to *The Times*, 13 December 1859, p. 5.
141. Worsaae, Letter to *The Athenaeum*, 31 December 1859, p. 890.

5

Consolidation

The Decade Begins 1860–3

The cork in the bottle labelled human antiquity had been pulled. The first impressions of the contents were positive. An agreed plan, executed as a field experiment, established Prestwich and Evans as expert witnesses who could determine the age of humanity. In their hands speculation became fact and Boucher de Perthes, once derided, was now elevated to the scientific pantheon. Joseph and John spoke with one voice from a shared script. The wine merchant reinforced the papermaker; the numismatist backed up the geologist. If they disagreed, they kept it quiet. But what they couldn't do was keep control of the time revolution they had kick-started in Amiens and Abbeville.

Three questions, in no particular order, drove the time revolution through the 1860s: the age of human antiquity, finding fossil evidence for humans, and allocating them a place in the nineteenth-century world. Joseph and John had a go at the first: *how old was old*? But they soon dismissed it as unscientific speculation. Others pushed back their tentative estimates of ten or twenty thousand years by a factor of five.[1] And one mathematician, the largely unheralded but nonetheless remarkable James Croll, made a breakthrough at the end of the 1860s that used climate change to tell time.[2]

The second question was raised by time revolutionaries and their opponents: *where were the human fossils*? There were mammoths, woolly rhinos, cave bears, and hippos galore from the caves and Drift deposits of England, France, and Belgium. But where were the people? The stone tools, accepted by most as genuine in 1860, formed a valid proxy for the missing evidence. Furthermore, the large numbers of these unpolished axes suggested that absence of evidence was not evidence of absence for the missing human fossils. Nevertheless, this lack of human remains was a 'Spectacle bizarre!'[3] according to the French geologist Jules Delanoue, speaking in Paris in February 1862.

A year later Jacques Boucher de Perthes believed he had found one of these elusive ancestors at Moulin-Quignon outside Abbeville. Prestwich and Evans combined with Falconer to act, once again, as expert witnesses. It pained them to say so, but this time they had to disagree. And while the deep disagreements about the Moulin-Quignon evidence continued, Thomas Huxley deftly supplied the sought-after evidence in *Man's Place in Nature* (1863) and passed it on in the same year to Charles Lyell to incorporate into his *Antiquity of Man*.

The final question stirred the nineteenth century's waters of colonialism, race, savage customs, and human progress: *what did they look like?* It was fine to have ancient stone implements and ancestral skeletons. But what did this say about their character and level of development? At a time of relentless progress, how could they be folded into the history of humankind? This task was seized on by John and Joseph's friend, John Lubbock who became Sir John when his father died in 1865. That year he published his bestseller *Pre-Historic Times*, in which he turned his natural history gaze on early humanity. In the same year the Oxford anthropologist Edward Burnett Tylor (1832–1917) brought out his influential *Researches into the Early History of Mankind*. Where Lubbock went from the archaeology to the living people, Tylor did the opposite.

Slipping Away

In the decade that followed the events of 1859 it was Prestwich who was most affected by what happened when others such as John Lubbock, Edward Tylor, Charles Lyell, and Thomas Huxley started to explore the possibilities of a deep human history that he had uncorked:. 'The more one becomes acquainted with time,' he wrote in March 1860 to Kate Thurburn, the second of his five sisters, 'the more slippery does it appear to be.'[4] Here was a fond, middle-aged older brother amazed at the speed with which birthdays came around and Kate's children were growing up. Another sense of slippery time would grow on him during the next four years, after which he gave up investigating the geology of the oldest humans. His clenching authority in these matters was troubled by two incidents, professional and personal. He became unsettled in 1863 by the scientific slights in Lyell's book *Antiquity of Man*. Falconer, enraged by Lyell's treatment of Prestwich, vehemently pressed the case as one of honour among gentlemen scientists. Two years later Hugh Falconer died. And if this body blow was not enough for the tightly wound Joseph to bear, it was followed a year later by the death of Civil Mary, his youngest and 'best of sisters',[5] his indispensable amanuensis and companion at their shared home.

The fight was knocked out of Prestwich the time revolutionary. After 1864 his only major publication on human antiquity appeared in 1873 with the long-awaited report on Brixham Cave. Only in old age and strengthened by Falconer's niece Grace McCall, whom he married in 1870, would he return to the topic. But this involved a disastrous endorsement of pseudostone artefacts found near their home in Kent.[6] In the 1860s he poured his geological energies instead into Royal Commissions on the nation's water supply[7] and coal deposits[8] and in 1874 examined the possibility of a Channel Tunnel.[9]

Meanwhile, Steadfast John ploughed steadily ahead. He did not give up on human antiquity or stone tools. He now became their greatest collector. Throughout the decade he published notes on what he was either sent or found, culminating in 1872 in his encyclopedic, copiously illustrated tome, *Ancient Stone Implements, Weapons and Ornaments of Great Britain*. By 1865 Prestwich would have nothing more to do with caves apart from ridding himself of the bothersome Brixham backlog. Evans, on the contrary, became a founder member of the excavation committee for Kent's Cavern, Torquay. Lyell was the committee's chairman and Pengelly an instrumental member. John Lubbock, who accused Lyell of plagiarism that year, was a member also.[10] Annual reports were written throughout the rest of the decade and beyond.[11]

John's scientific interests broadened during the 1860s to match his growing prosperity as a paper magnate. He was elected a fellow of the Royal Society in 1864, his certificate signed by Prestwich, Huxley, and Lubbock. He was increasingly found at the committee tables of the scientific establishment. Here he dispensed his interdisciplinary knowledge and supplied much-needed business acumen that kept several learned societies financially afloat. As befitted a treasurer, he regularly wrote papers during the decade about his first passion, the British coinage. Away from Joseph he lost some of his inhibitions about theory and scientific speculation. He put forward a theory about ice ages, complete with a working model, child's play for someone who invented an envelope machine, made from a wheel rotating in a box.[12] He delved back into the remote fossil past. 'Hail Prince of audacious Palaeontologists!', Falconer wrote to him on 15 January, 1863. 'I hear that you have today discovered the teeth and jaws of the *Archaeopteryx*. Tomorrow I expect to hear of your having found the liver and lights!'[13]

Two years later John's painstaking reinterpretation of the archetype for an evolutionary missing link, the half-bird, half-reptile fossil *Archaeopteryx* appeared in print.[14]

Both men, however, were eclipsed in this decade by their friend John Lubbock. By 1870 he had written a second bestseller, *The Origin of Civilisation*. In the same year he entered Parliament as Liberal MP for Maidstone, Kent. Lubbock was very clubbable, and his position in Darwin's circle, his wealth, and now his political career made him an indispensable ally for those like Huxley who saw evolution and deep human history as unfinished business.

Throughout the decade science made headlines while its controversies drove circulation. On both counts biological evolution was a newspaper editor's dream. But so too was great human antiquity, pronounced by Darwin as 'the most interesting subject which Geology has turned up for many a long year'.[15] Public controversy was part of the territory. Lyell's slight that Falconer and Prestwich felt so sharply was played out in the highbrow pages of *The Athenaeum*. Lubbock's grievance over Lyell's suspected plagiarism was a less public, but potentially more damaging spat. And before Prestwich quit the field, there was the widely reported scientific controversy in 1863 over the authenticity of that human jaw from Moulin-Quignon.

The decade in which the effects of 1859's time revolution were first felt was no longer the sun-kissed Victorian heyday which began with Prince Albert's declaration in March 1850 that:

> Nobody...who has paid any attention to the peculiar features of our present era, will doubt for a moment that we are living at a period of most wonderful transition, which tends rapidly to accomplish the great end, to which, indeed, all history points—the realization of the unity of mankind.[16]

Ten years later and the international mood was sombre, the unity of mankind a fading dream. War had returned to Europe with the latest phase of the Risorgimento in 1859. Two further conflicts followed; the Danish-German war of 1864, and a year later the seven-week war between Austria and Prussia. In 1861 the greatest continental conflict ever seen, the American Civil War[17] started as secession threatened to break up the Union over the institution of southern slavery. The decade closed with the Franco-Prussian War that began in 1870 and ended a year later in disaster for France.

Britain stayed out of major conflicts in Europe but fought incessantly to establish new colonies and suppress insurgencies. Britons had been profoundly shaken in 1857 by the rebellion in India[18] and the 1860s saw colonial wars and military action in China, New Zealand, West Africa, Japan, Jamaica, Afghanistan, Bhutan, and the Andaman Islands.[19] Added to these continual expenses on the Exchequer was the calamitous collapse in 1866 of a City of London bastion, the Overend Gurney Bank, a downfall that almost took the whole financial system with it.[20] The Age of Equipoise of the 1850s was replaced with the lost decade of the 1860s.[21]

Cherchez les fossils humains 1860–3

The decade began well for Evans. After much lobbying, some of it led by himself, the duty on paper was abolished in June 1861 during one of Gladstone's heroic budgets.[22] This simplified his life and enriched his business during some boom years for paper products. But much to Evans's annoyance, Gladstone left in place the duty on importing rags from France, a vital raw material for papermaking.[23] Prestwich had only reasons to be glum. New taxes affecting the wine trade added to his administrative burden. These solicited commiseration from Darwin in a letter asking for Joseph's views on *The Origin*.[24]

Towards the end of 1860 there were two further sceptical accounts of their time revolution in the Somme. John Henslow, Darwin's old master,[25] continued to question the association between the extinct animals and the stone artefacts.[26] Was it possible, he asked, to demonstrate that the beds of gravel which contained the bones and stones were 'undisturbed'? In his view rivers gathered together materials from different periods of time. The jumble of shells, animals, and stone tools in their gravels were never a pristine account, like the *in situ* photograph taken in the Fréville pit at Saint-Acheul by John and Joseph to convince those not present. Most of his criticism, however, was directed at Boucher de Perthes, whom he took to task for interpreting the river gravels as a widespread cataclysmic flood and for using unscientific concepts such as antediluvian and pre-Adamite. 'Prehistoric'[27] was Henslow's chosen scientific term, even though he disputed that the great age it implied should be applied to the flint implements.

A second, lengthy criticism came from the American-born geologist Henry Darwin Rogers (1808–66), who in 1857 became Regius Professor of Geology at Glasgow. Writing in October 1860 for the middlebrow audience of *Blackwood's Magazine*, he drew on his visit to the Somme in August 1859. A firm believer in Mosaic creation,[28] Rogers went to Abbeville, much as the evangelical writer Isabelle Duncan had done, to 'Go but not see'. Unlike her he did accept the stone tools as human artefacts. But on three further points he tried to shred Prestwich's geological case for their great antiquity. First, he argued, the juxtaposition of stones and bones did not necessarily indicate their contemporaneity. Shades of Henslow's river jumble here. Second, and spuriously, he claimed that the great age of the animal bones was unknown and doubtful *because of* their association with the flint implements. Uncertainty feeding off uncertainty. Finally, he maintained that as the age of the Drift sediments could not be quantified in thousands of years, the argument for the great antiquity of their contents must fall.[29] His final judgement on Prestwich's case was delivered in the Scottish legal idiom 'Not Proven'.[30]

Neither Joseph nor John replied, but John Lubbock took a swipe at him in 1862 in the short-lived *Natural History Review*.[31] From the heights of a highbrow journal Lubbock loftily set out to correct some middlebrow misconceptions about geological methods. Any geologist who had paid any 'special attention' to the subject of contemporaneity and age, he wrote, *must* agree with Prestwich. Then he went on: 'Fortunately, however, there is one exception to this rule'—Rogers' article in *Blackwood's Magazine*.[32] 'Fortunately' dripped sarcasm over Rogers' tortured logic before he drove home the point. He quoted verbatim Rogers' observation that a few minutes' inspection of the beds revealed they lacked fissures and were too compact to 'admit of any... insinuation or percolation of surface objects'.[33] Q.E.D. The flint implements and the animal bones really belonged to the same age as the sands and gravels where they were found.

Belt, Braces, and Muscular Scientists

What John and Joseph did instead was bury any and all opposition under a thick drift of facts. They achieved this with two lengthy follow-up papers for the Society of Antiquaries and the Royal Society. Two years after their visit to Amiens and Abbeville Evans got to his feet on 16 May 1861 to give an *Account of Some Further Discoveries of Flint Implements in the Drift on the Continent and in England*. His purpose was to show how common and widespread these unpolished stone implements had become in just two years. Now the Seine, Ouse, and Thames were added as happy hunting grounds for oval and pointed axes and flint flakes. These find-spots made it possible to repeat the original field experiment that confirmed human antiquity. For good measure he provided a roll call of all those who followed his advice to 'Go and see' and who were convinced by the archaeology and geology.

Among the many new finds he was particularly excited by were two fine flint axes from Reculver, near Herne Bay in north Kent. These came from a different context, a sea cliff rather than a river gravel.[34] Every year the collapsing cliff brought down more water-laid drift deposits that formed its capping. As a result flint artefacts could be picked up on the beach below. The find at Reculver led to John's irrepressible good humour on the day his wife Fanny was caring for a dying cook at Nash House.[35] Two years later in 1863 and the axes still had pride of place as full-size engravings when they were included in his *Archaeologia* paper. In addition, a fold-out plate by the engraver Joseph Lowry (Figure 5.1) illustrated twenty further implements: flakes, and pointed and oval shaped axes. It set out the likely range of variation for these implements and so became the template for classifying all subsequent unpolished artefacts of great human antiquity.

Joseph read his long paper with a very long title *Theoretical Considerations on the Condition under which the Drift Deposits Containing the Remains of Extinct Mammalia and Flint Implements Were Accumulated, and on their Geological Age* at the Royal Society on 27 March 1862.[36] When it was published in the *Transactions* in 1864, he combined it with another paper *On the Loess of the Valleys of the South of England, and of the Somme and Seine.*[37]

The paper was produced under duress, which shows in the original manuscript in the Royal Society Library. Every page is heavily annotated, passages crossed out, rewritten, and cut and pasted. Compared to his 1860 manuscript announcing the discovery, it is a dog's breakfast, the product of a wavering mind. And to add to his dithering was a highly negative referee's report from the Cambridge mathematician and geologist William Hopkins (1793–1866). His principle objection, written in April 1862, concerned Joseph's 'speculative discussions' about the climate when the beds of gravel were laid down. The editor Sir Andrew Ramsay, another geologist, sat on this report, and then farmed it out a year later to Lyell, who in the briefest of notes declared it should be published in its present form. And it was.[38]

Here was the heart of the geological matter surrounding human antiquity, rivers, and ice. Joseph firmly believed in a single ice age. Within this framework he explained the Drift geology of the Somme as the result of

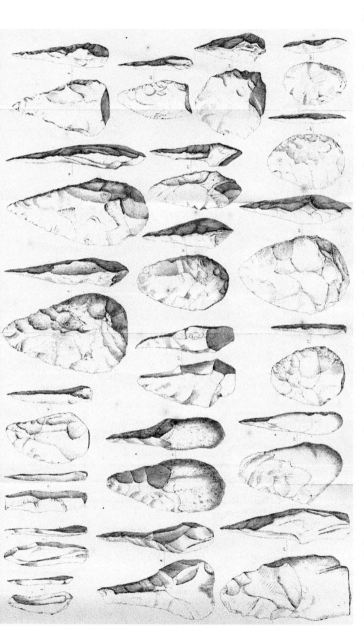

Figure 5.1 Evans used these twenty stone implements to show what by 1863 he regarded as typical ancient stone artefacts. Oval (11, 16–19) and pointed (5–10) forms dominated amongst the finished axes, and subsequent research has confirmed this basic classification. Unlike his much earlier coin sequence (Figure 3.4), the stone implements were not presented as a developmental drawing, with one form leading to another, but as distinct end products of an intelligent manufacturing process.

Source: Evans 1863: Plate IV.

rivers which, during the Pleistocene, were much larger. Intermittent floods after the annual snow melt spread sheets of coarse gravels across the valleys, and fine silts, known as loess, were also deposited.[39] Large boulders in the same gravels, Joseph argued, had been moved by ice when the rivers froze.[40] A cold glacial climate for gravels and loess was indicated by the extinct woolly mammoth and rhino. And these were not the only ice age animals. John Lubbock, accompanied by the Rev Charles Kingsley, had unearthed in 1855 the skull of a musk ox, an extant high Arctic species, close to the engine house at Maidenhead railway station.[41] Joseph's final point, which did not change over the next forty years, was that the deep incision of valleys like the Somme was due to the rise of the coasts and the sinking of the land in southern England and northern France. The change in altitude increased the erosive power of rivers creating the modern landscape.[42] This was Hopkins's major complaint of unsubstantiated speculation.

But Joseph's frequent field trips had paid off. What in 1859 had been a singular geological observation could now be presented as a general rule. In all the valleys of France and England which Prestwich visited he found that height above the valley floor distinguished two geological deposits and their archaeology (Figure 5.2).[43]

The high-level sands and gravels contained Boucher de Perthes's unpolished oval and pointed axes together with extinct animals. Menchecourt at Abbeville and the pits at Saint-Acheul at Amiens would slot in here. In the low gravels, Prestwich pointed out, simple flakes predominated among the implements.[44] Flakes, in his opinion, were for cutting and flaying carcasses, pointed weapons for defence and offence, while some spatula-shaped pieces could, he believed, have acted as ice chisels to enable fishing.[45]

As Lubbock perceptively pointed out, Prestwich's belief in a single ice age required no repeated changes in climate.[46] An age of ice was simply

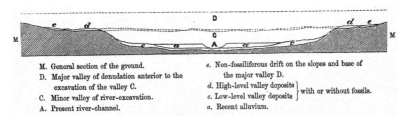

M. General section of the ground.
D. Major valley of denudation anterior to the excavation of the valley C.
C. Minor valley of river-excavation.
A. Present river-channel.

e. Non-fossiliferous drift on the slopes and base of the major valley D.
d. High-level valley deposits ⎫ with or without fossils.
c. Low-level valley deposits ⎭
a. Recent alluvium.

Figure 5.2 Prestwich's cross section of a valley and its two levels of gravels, high (*d*) and low (*c*). The loess deposits followed a similar division.
Source: Prestwich 1864.

succeeded by the present age of warmth: the Post-Pliocene (Pleistocene) became the Recent (Holocene) geological period. But as early as 1862 in his paper on the formation of the Somme river terraces Lubbock hinted at a more complex picture. The problem was those hippos in the lower valley gravels. These indicated warm conditions, raising the possibility that there was not a single undifferentiated ice age but rather a cycle between cold, warm, and cold stages. Prestwich would have none of it. If you can have woolly mammoths and rhinos adapted to the cold, you can have also woolly hippos.[47]

What Lubbock did was refine Prestwich's account of how small pockets of gravel, sand, and loess clinging to the valley sides could be turned into a deep geological history (Figure 5.3). He described them as a series of terraces, rather than simply two, as Prestwich did, implying a staircase of descending age, oldest to youngest, carved by the rivers into its chalk sides.

Lubbock was a recent convert to the wonders and power of ice and glaciers. He became a frequent visitor to the Alps to see the Neolithic and Bronze Age lake villages clustered around the major lakes. His first venture into the mountains was with his friend, the Irish physicist John Tyndall (1820–93). Together with Thomas Huxley, the three scientists crossed the Rhône glacier in July 1862.[48] Then, leaving Huxley behind, they set out to ascend the Jungfrau. Tyndall recalled that to Lubbock 'the scene opened out with the freshness of a new revelation'.[49] Soon, however, he was wishing himself elsewhere, as the pair

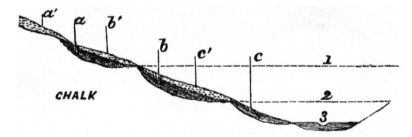

Figure 5.3 Lubbock's interpretation of the Somme river terraces. The higher the terrace, the older its contents. And within each terrace the gravels (a,b,c) are older than the loess (a',b',c') above. This diagram is close to our current understanding of river terraces formed under successive cold and warm conditions (Bridgland 2014). Terraces form as environmental changes alter the rate at which rivers downcut their valleys. But any uplift of the land further accelerates the downcutting power of the rivers and results in a steep valley such as the Somme

Source: Lubbock 1862.

had to rescue, with great difficulty, one of their overenthusiastic porters, who had fallen into a crevasse (Figure 5.4). When he left to fetch help after the successful rescue, Tyndall approvingly noticed 'moisture in Lubbock's eye. Such an occasion brings out a man's feeling if he have any.'[50]

Muscular is an adjective usually applied to the 1860s Christianity of the Rev. Charles Kingsley and the ethos of the English public school.[51] But as this incident shows, it could equally be applied to the 1860s breed of scientist such as Tyndall and Huxley. Tyndall was among the most muscular of the breed. In the summer of 1859 he spent a night at the top of Mont Blanc with the chemist Edward Frankland (1825–99) conducting scientific experiments.[52] For his part, Prestwich enjoyed an energetic climb but never attempted any great ascent.[53]

By the time the two Johns and Joseph had prepared and published their papers, one of their doubters, Henslow, had died. A critic like Rogers would never be convinced. Joseph had no time for his Scripture-based geology or Boucher de Perthes's argument for a cataclysmic flood to explain the Abbeville gravels. But in a significant passage at the end of his 1864 *Transactions* paper— the last major work he would write on the Somme—he stated his personal belief that, having looked in detail at the great changes to the earth through its recent geology, 'I confess I feel deeply and strongly impressed with the probability that in this unexpected succession of changes we may trace evidence of great and all-wise design.'[54] And in a textbook example of nineteenth-century geology at its most practical the purpose of the all-seeing, divine architect was to give stability to the earth's crust so that human civilization could flourish.

Prestwich was on a different track from that of his close friends, Lubbock the evolutionist and Evans the businesslike scientist.[55] While religious sentiments never drove a rift between them, it nonetheless defined different boundaries to their scientific lives. These limits aligned Joseph closely with Falconer, who kept his faith separate from his geology. Writing to Grace McCall in 1854 before he returned to London, he set out his philosophy of two spheres: 'The Almighty has given us reason,' he instructed his young niece, 'to investigate the laws and order of creation... As regards the creation of the world, the evidence is as clear that millions and millions of years must have elapsed between the first appearance of life on the earth and the present day.'[56] And yet he reminded her in the same letter that 'what I have said here bears solely upon our knowledge of the physical world, and not upon doctrines of faith for our moral and religious guidance'.[57] But since 'true religion and true science can never be irreconcilable', their task was to bring together what others, like Huxley, could not.[58]

WHYMPER SC.

Figure 5.4 'Recovery of Our Porter'. According to Tyndall's account, Lubbock was holding the rope, which was made of coats, waistcoats, and braces because the climbing rope had fallen into the crevasse with the porter. Apparently, 'Lubbock's waistcoat proved too tender for the strain.' Lubbock later described his Alpine excursion to Darwin as 'capital fun'.

Source: Tyndall 1871.

Grace cared for and supported both these men. She moved into the Falconers' London house in Park Crescent[59] in 1861 and for four years she managed both Uncle Hugh and the household. Five years after Falconer's death she married Joseph. With her religious convictions and scientific knowledge she buttressed both men's insistence on separate spheres. Prestwich might have refused to intone the Athanasian Creed in church,[60] too Catholic for his taste, but he remained troubled if geology impinged on his Christian faith. At the end of his life, and with Grace's approval, Joseph reiterated the separate spheres approach that kept his science and religion apart. In response to Arthur H. Tabrum, who was collecting letters for his book *Religious Beliefs of Scientists*,[61] Prestwich wrote, 'Religion and science constitute two distinct branches of human knowledge and inquiry. They move in parallel lines, and cannot, in my opinion, clash. They certainly should not. The one has to deal with moral questions, the other with physical questions.'[62]

Evans saw it differently. So long as religion did not impede the progress of scientific enquiry, then he did not feel it was a vital issue.[63] This was Lubbock's view also. He suffered no crisis of faith and lived contentedly with the intellectual marriage of science and religion, writing that 'men of science and not the clergy only, are ministers of religion'.[64]

This did not make Evans and Lubbock agnostics like Huxley, the type fossil of doubt. They attended church and were buried with full religious honours. Neither did it put them through the emotional wringer which Lyell experienced as he sought to balance his heartfelt beliefs with the evidence of his geological eyes.[65] For Lyell it was a generational thing: Lubbock, being younger, had, 'comparatively little to abandon of old and cherished ideas'.[66] Keeping the spheres separate would always be a challenge once humans became part of the geological story. It was the time revolution, initiated by Prestwich and endorsed by Lyell, that broke the compromise which for scientists like these two kept theological arguments away from geological facts.[67] By filling the 'big ditch', they had created a pitfall for their own faith.[68]

Developmental Drawings and Evolutionary Pictures

The beginning of the decade saw Huxley's rapid rise as *the* public intellectual championing the cause of evolutionary change, a role announced by a barnstorming riposte to Bishop Wilberforce at a meeting of the British Association for the Advancement of Science chaired by Henslow and held in Oxford on 30 June 1860. Huxley won the audience by expressing his preference for a miserable ape as an ancestor rather than the clergyman opposite him who ridiculed serious science.[69]

A year later came his acrimonious disagreement with Richard Owen (1804–92), London's leading anatomist, palaeontologist, and critic of natural selection.[70] They clashed repeatedly over the brains of apes and humans.[71] Huxley saw continuity in their structures. Owen regarded the human brain as unrelated to any primate cousin. Huxley's comparative anatomy proved correct. But Owen refused to concede, which resulted in Huxley's disdainful put-down, 'Life is too short to occupy oneself with the slaying of the slain more than once.'

Such feisty exchanges attracted satire, and a poem *Monkeyana* written for *Punch* duly appeared in 1861 (Figure 5.5).[72] Six of its thirteen verses lampooned the spat between Owen and Huxley over the brain, while two were devoted to human antiquity, naming Prestwich and Pengelly, much to the latter's amusement[73]

Monkeyana's gorilla is a distressing image today because of its inescapable racist references. These did not much concern the Victorians. But it was disturbing in another way to Mr Punch's readers and anathema to a scientist like Richard Owen, who regarded human creation as evidence of a divine hand. Four legs walked on instinct, two legs on reason.[74] Apes and humans had to be kept in their own separate spheres or else moral chaos ensued. Age was important to maintain the boundaries. The longer the chronology, the more chance there was that biological evolution became inevitable under Darwin's argument of unlimited time—a position neatly summed up in one of *Monkeyana*'s better verses:

> Let pigeons and doves
> Select their own loves
> And grant them a million of ages,
> Then doubtless you'll find
> They've altered their kind,
> And changed into prophets and sages

The unthinkable was that gorillas could also turn into humans. Soon, humourists like Charles Henry Bennett (1828–67) were turning everything

Figure 5.5 *Monkeyana* appeared in *Punch* on 18 May 1861 and is better known for the arresting image of a staff-holding, sandwiched-boarded gorilla than its poorly scanned verses. The image is an unsubtle parody of the famous Wedgewood abolitionist medallion depicting a chained and kneeling slave asking, 'Am I *not* a man and a brother?'
Source: *Punch* 1861.

Then there's PENGELLY
Who next will tell ye
That he and his colleagues of late
Find celts and shaped stones
Mixed up with cave bones
Of contemporaneous date.

Then PRESTWICH, he pelts
With hammers and celts
All who do not believe his relation,
That the tools he exhumes
From gravelly tombs
Date before the Mosaic creation

into something else through what he called 'developmental drawings' (Figure 5.6).[75] Time was proving slippery indeed.

Back in the world of science, Lubbock told his readers in *The Natural History Review* they should not be surprised at the lack of human bones in the Somme: 'No animal as small as a man' had been found in any of the rivers so far searched;[76] small bones had perished leaving behind only the robust skeletons of rhinos and mammoths and stone tools. He had a point. But equally, the methods of digging with pick and shovel and the lack of sieving through a fine mesh would skew recovery to the large mammals. But John's conclusion remains irrefutable, 'negative evidence in palaeontology must . . . always be regarded with suspicion'.[77]

The *Monkeyana* verses about Prestwich and Pengelly brought to the fore a fact that Lubbock could not know but archaeologists have since discovered: the chances of finding human fossils are higher in a cave than in a gravel terrace. In 1859 the caves Pengelly was digging, Kent's Cavern and Brixham, and Hugh Falconer's Grotta Maccagnone in Sicily all had stone tools and extinct animals. But none of them had yet yielded up human fossils.[78] But elsewhere human skulls had been found in caves, notably the Engis Caverns near Liège in Belgium (1829), Forbes' Quarry on Gibraltar (1848), and the Feldhofer Grotte in Germany's Neanderthal (1856).[79] And in 1863 a collaboration between the French palaeolontologist Édouard Lartet (1801–71) and the English businessman and antiquary Henry Christy (1810–65) had excavated much of a skeleton in the limestone rock shelter of La Madeleine by the banks of the Vézère in the Dordogne.[80] Limestone is good parent rock for bone preservation. Caves are restricted spaces that

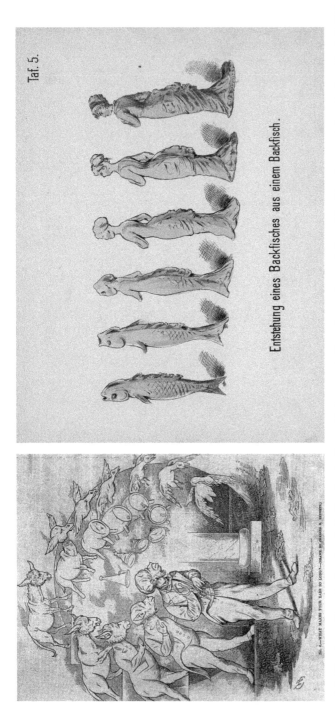

Taf. 5.

Entstehung eines Backfisches aus einem Backfisch.

Figure 5.6 Thanks to Darwin the decade held a fascination with transmutation, change by evolution. The 'development drawing' on the left, *What Makes Your Ears So Long?*, by the English humourist Charles Bennett first appeared in 1863. It shows a dishevelled sage turning into a goose via an ass, which also evolves into a barrel and a dunce's hat. It could be a graphic comment on the verse from *Monkeyana* about prophets and pigeons. (Bennett 1863). On the right is one of twenty coloured images by Franz Schmidt published in Germany during the 1860s. It shows the evolution of a fried fish (*Backfisch*) into a teenage girl, who, in the slang of the time, was also known as a *Backfisch* The direction of the figures suggests the process might also be reversed. Evans's chart of the British coinage (Figure 3.4) is an antiquary's example of a developmental drawing.
Source Bennett 1863 [1872]; Wellcome Collection.

concentrate finds and focus even haphazard excavators on their contents in ways that an expansive open gravel pit cannot.

In the last 150 years the caves of Europe have yielded up a rich collection of human fossils.[81] By comparison gravel pits, ancient beaches, and lake sites such as Hoxne have hardly produced anything. Some of the earliest human burials in Europe come from caves, as at La Chapelle-aux-Saints, France, where a complete Neanderthal skeleton was found in 1908. The best open locales can manage is a chewed shin bone from the otherwise perfectly preserved landscape of Boxgrove, southern England[82] and a robust jaw from Mauer outside Heidelberg, Germany, discovered in 1907.[83] In 1860 Lubbock could not know that by searching the Somme terraces the time revolutionaries were looking in the wrong place to find the fossil remains of the ancestors who made those unpolished stone implements.

John Lubbock did hope in July 1862 that Boucher de Perthes would 'be the fortunate finder of the first human bones in the drift'. Six months later and 'the veteran antiquary of Abbeville' startled the scientific world, again, with a claim that he had done just that.

Two Books Seal the Revolution

The second great year in the nineteenth century's time revolution was 1863. But it was also one of its most acrimonious. It began with two major books, Lyell's *Antiquity of Man* and Huxley's *Man's Place in Nature*, published respectively on 6 and 20 February. Huxley's short book was greeted with glee by Darwin: 'Hurrah the monkey book has come.'[84] Delight was muted for Lyell's much longer volume, weighing in at over 500 pages and proudly sporting an embossed mammoth tooth and Amiens flint axe on its cover. The general feeling was that Lyell had bottled it when he came to discuss change brought about through Darwin's mechanism of natural selection. On the last page of the last chapter, he wrote: 'The whole course of nature may be the material embodiment of a preconcerted arrangement' and, if so, 'the perpetual adaptation of the organic world to new conditions leaves the argument in favour of design, and therefore of a designer, as valid as ever'.[85] From an evolutionary viewpoint human history was, in his closing words, 'the ever-increasing dominion of mind over matter'—doves designed by sages.

Lyell's sensibility to old and cherished beliefs had won out over his rational geological mind. But there were fewer complaints from the biologists about his chapters on the geological age of humans. He took the reader on a lengthy promenade around the discoveries of the previous four years and earlier to build an unassailable case for human antiquity. Here was the geological grandmaster claiming the field of human antiquity as his own. As before, he offered no estimates of the ages involved but clearly now favoured a long chronology. The only hint as to how long was long came the following year, when he addressed the British Association for the Advancement of Science at its annual meeting in Bath. Hidden away at the end of his review of the year in Geology was a comment about how old the most recent geological period, the glacial and postglacial, might be: 'When called upon to make grants of thousands of centuries in order to explain the events of what is called the modern period, [geologists] shrink naturally at first from making what seems so lavish an expenditure of past time.'[86] His message was to 'make more liberal grants of time to the geologist'. He was clear in his demand that geologists cast off the fetters of 'old traditional beliefs', the 6,000-year-long history for humans and everything else in Mosaic creation. But whether 'thousands of centuries', 100,000 years, was his preferred timescale remained opaque.

Not so for Lubbock, who seized on a statement by Lyell comparing the rates of sedimentation in the Mississippi and the Somme. Were fossil bones in the former older than the stone implements in the latter? To which Lyell gave the answer no,[87] and which Lubbock interpreted as Lyell's opinion that 'the drift hatchets of England and France are at least 100,000 years old.'[88] Look hard enough and you will find the answer you want, but scarcely a bold, unequivocal statement of great age.

The monkey book did slay the slain once again: Huxley could not resist recounting Owen's mistakes concerning the gorilla's brain.[89] It tickled Darwin's bulldog that at the same time Evans had found the jaws and brain cavity of *Archaeopteryx* preserved in its limestone slab, which Owen had been poring over in the British Museum only to miss this vital evidence.[90] To compound Owen's embarrassment, Evans used his discovery to make for him a rare comment on *Archaeopteryx*'s significance 'for the great question of The Origin of Species':[91] a reptile with feathers and a bird with teeth. Evolution was proving more inventive than the comic developmental drawings of sages and geese, fish and young women.

Man's Place in Nature provided the first fossil ancestor who might have made the unpolished stone implements, the skull from Neanderthal (Figure 5.7). Courtesy of his close friend the palaeontologist and surgeon

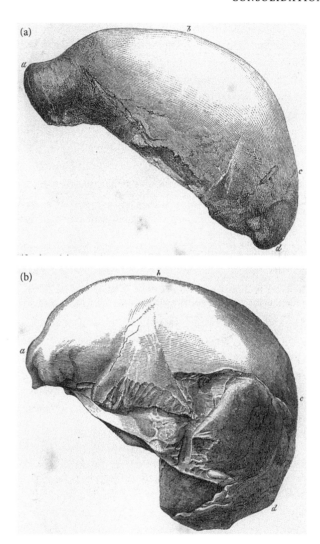

Figure 5.7 (a) Huxley's side view of the cast of the Neanderthal skull. The robust brow ridge is to the left.

Figure 5.7 (b) The cast of the Engis skull with its high forehead and reduced brow ridge to the left.

Source: Lyell 1863.

George Busk (1807–86), Huxley had a cast to work with. And he also relied heavily on a detailed account, using Busk's translation,[92] by Professor Hermann Schaaffhausen (1816–93), an anatomist at the University of Bonn.[93] Huxley did not regard the Neanderthal skull as a separate species but as an extreme variant of *Homo sapiens*.[94] He shared this analysis with Lyell, who included it in *Antiquity of Man*.

Huxley overlaid the profiles of three skulls, like three hats in a developmental drawing: a modern European on top of the Neanderthal fossil above the Chimpanzee (Figure 5.8).[95] The sequence showed continuous development in the height of the braincase and hence the size of the brain it contained. And in a second profile the three skulls started with Neanderthal followed by a modern Australian and crowned by the Engis fossil from Europe.[96]

This development of skull shape could be read several ways: a gradual change resulting from a continuous process of natural selection as Darwinians like Huxley would favour; or a justification of Spencer's law of progress that simple always becomes complex, and where measuring the convolutions and the size of the brains these skulls contained would make his point. What these diagrams set out to disprove was the 'big ditch' favoured by scientists like Owen that separated apes from humans. Falconer, Grace McCall, Prestwich, and Lyell would have been uneasy that by filling in the ditch they had made Huxley's case more likely.

Huxley used these skulls as the first signposts of an evolutionary journey that he expected to grow in complexity as more specimens were found and the range of anatomical variation broadened. He placed Chimpanzee as the primate ancestor, a proxy for the oldest. Above them was the Neanderthal skull of very great antiquity with marked differences from a modern skull. Top of the pile was Engis, which was of great antiquity, closely equivalent anatomically to his own skull. And from a Victorian's perspective the position of the Engis fossil skull *above* a modern Australian met their expectations derived from a racial geography that placed 'savage' hunters below 'civilized' farmers and Europe above any other continent.

The outlier skull was the big-brained one from Neanderthal.[97] It had a characteristic long, low vault, big eye sockets surmounted by immensely thick brow ridges, above which a forehead of only middlebrow proportions rose. Huxley concluded this was not an advanced ape but a cold-adapted ancestor, the first of the cavemen.[98]

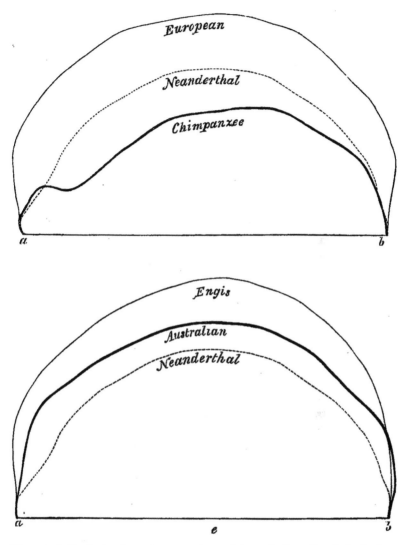

Figure 5.8 Huxley's schematic comparison of three skull profiles. In both, the front of the skulls is to the left.

Source: Lyell 1863.

Letters to the Papers

How did the time revolutionaries respond to these two books? Approval for Huxley's from Lubbock but disappointment about Lyell's muted endorsement for his old mentor, Darwin and *The Origin*.[99] Silence from Evans and Prestwich. Rage from Falconer. Uncle Hugh was incensed, firing off a letter, 'Primeval Man—What Led to the Question?', published in *The Athenaeum* on 4 April. Lyell's crime was to claim he started the bandwagon rolling when in fact he had jumped on it:

> Sir Charles Lyell struck in with the set of the current, to take up the question when it was launched as a proved case in 1859, and while Mr Prestwich and myself were still occupied in following up our enquiries. This alone should have made him scrupulously careful in his statements.[100]

Over five tightly packed columns Falconer laid out what he saw as false claims to originality and slack reporting of facts. The charge was sloppy science bordering on the ungentlemanly. None of his criticisms dealt with the contentious issues of biological change or even Lyell's elevated estimate for the Drift deposits.

Falconer wrote with Joseph's approval and Lubbock's support.[101] The reply from Lyell was slow in coming. He had been out of town so that his reply to Falconer's 'formal complaint' appeared on 18 April. The tone was weary and aggrieved. Quite frankly he couldn't see what the fuss was about. He had cited Prestwich's work on the Drift in *Antiquity of Man* and praised it. He gave similar recognition to Falconer's work at Brixham. He took umbrage at Falconer's principal objection that, together with Prestwich, 'they were the first to settle the true chronological relations of the glacial period and certain fluviatile drifts and cave deposits in which the bones of extinct quadrupeds, and in some of them also human remains occur'.[102] Not so. Lyell traced his involvement back to his groundbreaking *Principles of Geology* of 1835, written before Falconer and Prestwich began their geological careers, as well as intoning the names of French geologists who had determined the question of human contemporaneity with extinct animals 'long ago, although they propounded their truths to an unbelieving generation'.[103] Falconer and Prestwich did not, he insisted, convert him to the cause of great human antiquity. But they did play a role in reforming the date for the most intense period of glaciation in Western Europe and, as a consequence, the relative age of humans.[104]

A week later it was Joseph's turn. Mild in comparison to Falconer, he nonetheless complained about the need for accurate reference to his work— scientific work which, and here his resentment of Lyell's privilege of being a full-time geologist of leisure boiled over, experienced delays because of the 'limited measures of time I can take from active business avocations'.[105] He reiterated his opinion, contrary to Lyell's, that what he had found brought the animals forward in time to after the glacial period rather than pushing humans far back in time to before the ice age.

No reply this time from Lyell. But Uncle Hugh would not let it go, even though his friends regarded his attacks as unnecessary and spiteful behaviour.[106] In Hooker's curt opinion, he had revealed himself as 'one of the 2 classes of Scotchmen, "Scotsman"—& "d—d Scotsman".[107] Lyell, after all, was one of the good guys and not to be treated with contempt, as Huxley did with the unsympathetic Owen. Falconer knew this before he first wrote to *The Athenaeum*, confiding in Grace, 'I must keep my temper. What a grumpy old Uncle you have got! Like an infuriated Toro... driving his horns into the porcelain ware of Sir C. L. [Charles Lyell]!'[108]

But he could not stop himself. Uncle Hugh's final letter on the matter on 2 May doubled down on his feelings of unfairness. He hurled Lyell's points back at him: 'It was only after June, 1859, that Sir Charles Lyell's conversion was declared. Mr Prestwich's researches and my own may have had little effect in bringing about the change, but they certainly immediately preceded it.'[109] And he ended with the roar of the overlooked: 'The day has gone by, when scientific works could be written in the style of Louis Quartorze: *La Géologie; c'est moi!—l'Ancienneté d'Homme; c'est moi aussi!*'[110]

Lyell's sin was scientific greed, infringing Prestwich's achievement by invoking his seniority and greater standing. But unlike his accusers he did write the bestselling book that consolidated the great age for human history. Neither did he bear a grudge, nodding Joseph's tortured paper through the Royal Society, when he could so easily have rejected it.[111] All Lyell did to the criticisms against *Antiquity of Man* was make a few minor corrections to the two further editions in April and November that same year. But by then Falconer's geological attention was directed elsewhere.[112] Boucher de Perthes had, as Lubbock hoped, found human remains at Abbeville.

The Affair of Moulin Quignon

The announcement came on 9 April in the local newspaper *L'Abbevillois*. The *terrassiers* at Moulin-Quignon, high on the chalk hills to the east of

Abbeville, had found not just stone implements but teeth and a human jaw. Nine days later it was reported in *The Athenaeum*, and on the same page as Lyell's reply to Falconer, by the vice president of the Royal Society William Carpenter (1813–85). And so began *l'affaire de Moulin-Quignon*.[113]

It had all started so well for Jacques. A *terrassier*, Halattre, had come to see him on 23 March with news of flint tools and a human tooth from a pit at Moulin-Quignon in the heart of Prestwich's high Somme terrace. Jacques visited and told two other diggers, Dingeon and Vasseur to investigate further. On 28 March they reported another tooth. Back to the pit went Boucher de Perthes with the antiquary and draughstman Oswald Dimpre (1819–1906), and in front of seven witnesses a human jaw, a broken flint axe, and another tooth, were carefully dug out of the gravels, with a pickaxe.[114]

Unlike April 1859, no photographs were taken and an announcement in the local press was preferred to a presentation before the learned societies of Paris. The slights of previous years still rankled. Jacques chose to spread the news personally to his supporters, the most important of whom was Édouard Lartet. One of the others he wrote to, the abbé Louis Bourgeois (1819–78) passed the information on to Prestwich and Evans, who happened to be in France studying the cliffs of the Loire.

The time revolutionaries of 1859 were reunited in Abbeville on 11 April. Falconer was in Paris and arrived at the Tête de Bœuf after John and Joseph had left on 16 April. He found a disconsolate Boucher de Perthes. As it was described by Alfred Tylor (1824–84), another geologist travelling with John and Joseph, Evans was immediately suspicious of the authenticity of the stone tools he was shown from Moulin-Quignon.[115] They had been artificially stained to make it look as if they came from the black seam of gravels. A quick wash revealed their true colours. John also noticed piles of recently knapped flints around the quarry. The *terrassiers* had perfected their skills to reap the rewards that Jacques offered for new specimens and, since 1859, they had been knapping to meet the demand from the steady stream of geological tourists to the Somme. Henry Christy, a skilled flint expert, was well aware that they used the long winter evenings to make a fresh batch for the arrival of visitors in the spring.[116]

Jacques was most put out. He told the Englishmen, as it was reported by Tylor, that he had pulled the jawbone out of the black seam of flinty gravel *in front of numerous witnesses*. Moreover, the three *terrassiers* were persons of irreproachable character. Why would they cheat, he asked? The going rate for a genuine flint axe was twenty-five to fifty centimes, one franc for a fine one,[117] and this low price would be depressed further if they flooded the market with replicas. He had once offered twenty francs if the *terrassiers* could produce a facsimile axe, and no one had managed to do it.[118] It did not occur to him that they might have wanted to disguise their proficiency in flint knapping for fear of cutting off a lucrative sideline.

Falconer, the palaeontologist, was more concerned with the jaw. Animal bones were worth much more than stone tools, and the price put on human remains would have been riches for the workmen. Initially he believed in the authenticity of the discovery.[119] On first inspection the jaw had some novel features that he wanted to compare with a large collection such as Huxley's at the Royal College of Surgeons in London. He wrote from Abbeville to Prestwich saying he thought it was possibly from an unknown European race. It needed more study to be sure.[120] The jaw, already a prized relic, stayed in France. But he was given permission to bring a tooth back for further study.

But before he could do this, and before Prestwich was able to undertake more geological work on the age of the pits at Moulin-Quignon, Carpenter spoke to the Royal Society. He then broadcast the news of a 'remarkable discovery' in *The Athenaeum* on 18 April, at times gushing—'M. Boucher de Perthes had yesterday the kindness to place in my hands this precious fragment'— although he had not been present—'I cannot myself conceive that anyone who carefully examines the undisturbed condition of the bed can entertain a doubt that the bone in question is a *true fossil*, dating back to the time of its original deposition.'[121]

Grumpy Uncle Hugh now added Boucher de Perthes to his list of targets. In a letter to *The Times* on 25 April he questioned not only the flint tools but *also* the jaw.[122] His judgement now pitted him against Armand de Quatrefages (1810–92), the leading French zoologist and anatomical expert in the Académie des Sciences. Both men commented on some unusual features, in particular the posterior angle of the jaw—what Falconer, with apologies for jargon to his *Times* readers, called 'a marsupial amount of inversion',[123] and with Busk's help, and Huxley's collection of skulls from around the world, he matched it with an Australian jaw. But another fellow of the Royal Society, his friend the dental surgeon, Sir John Tomes (1815–95),[124] went one better, producing a sackful of lower jaws from an old London churchyard that also had this distinctive 'marsupial inversion'. Emboldened, Uncle Hugh took a saw to the tooth he had brought back and found it was white and full of gelatine. This was a recent and not a fossil tooth: 'The inference I draw from these facts is that a very clever imposition has been practised by the *terrassiers* of the Abbeville gravel pits—so cunningly clever that it could not have been surpassed by a committee of anthropologists enacting a practical joke.'[125] Quickly, however, he concluded that none of this malarkey discredited the time revolution of four years previously.

The letter did not go down well in Abbeville and Paris. Falconer's private concern that it would be treated as an example of perfidious Albion traducing the French proved correct. Four years to the day since the proof of deep human history at Saint-Acheul, Uncle Hugh received Jacques's very long and in places hysterical letter denying all the allegations against the workmen and himself. The letter finished with vintage Boucher de Perthes:

> An eye witness writes to me, with all seriousness, that during a spiritualist seance which took place last Thursday, 23rd April, the fossil man [of Abbeville] declared that the disaster which caused his death, twenty thousand years ago, crushed him between two stones before depositing his jaw where I found it, and that his name was Yoé etc... The deceased antediluvian did not seem to bear a grudge against me.[126]

Uncle Hugh despaired for the old man and was losing sleep about the effect his criticisms were having on 'Poor dear Boucher'. He confided to his niece in London, 'Oh that you were here as a buffer to toil through the *acreage* of his compositions, and to sustain the pang of his wounded feelings!'[127]

The tone from Paris was measured but firm. Four years had seen a sea change in opinion, and the French scientific community led by Lartet and Quatrefages rallied to Boucher de Perthes support. Heels were dug in on either side of the Channel: 'The French savants, the more they went into the case, were the more convinced of the soundness of their conclusions, while their English opponents, the more they weighed the evidence before them, were the more strengthened in their doubts.'[128] Carpenter, for one, had changed his mind about the authenticity of the jaw since splashing it in *The Athenaeum*.

To cut through the impasse and protect Anglo-French scientific collaborations, a joint commission was agreed and convened in Paris on 9 May. It lasted five days. The nominated English deputies were Prestwich, Falconer, Carpenter, and Busk. Prestwich vacillated but eventually did appear much to Falconer's relief. Carpenter deferred to the others' judgement and left early.

On their side the French fielded, Quatrefages, Lartet, and the geologists Achille Delesse (1817–81) and Jules Desnoyer (1800–87), who had introduced Falconer to Boucher de Perthes in 1856.[129] The commission's president was the zoologist Henri Milne-Edwards (1800–85), well chosen because he was half-English, the son of a colonist from Jamaica, and a professor at the Muséum national d'Histoire naturelle of the Jardin des Plantes, Paris.[130] Boucher de Perthes was represented by l'abbé Bourgeois and Charles-Joseph Buteux, known for his work on the Somme Drift deposits.[131] Albert Gaudry, who had confirmed Prestwich's findings at Saint-Acheul in 1859,[132] was also in attendance.

The commission's deliberations were protracted. The first meeting lasted six hours and went over, in laborious detail, the features that distinguished genuine flint implements from modern imitations. It was a pity that Flint Evans, though asked, could not attend.[133] Similar careful attention was paid to the jaw and the geology. There were moments of drama as Busk sawed into the jaw, revealing how easily the black staining on the outside washed off, while inside its cavity they found sand that did not match the Moulin-Quignon sediments.[134] There were moments of farce when Quatrefages claimed that Boucher de Perthes had muddled up the human finds and put them in the wrong boxes. The tooth given to Falconer when he visited and subsequently analysed in London should be discounted.[135] And anyway, this tooth did not come from Moulin-Quignon but Menchecourt on the other side of the town.

For the next three days the conference that had become a cause célèbre in Paris dragged on. Then, unannounced, the commission upped sticks and descended on Abbeville on the morning of 12 May. *Terrassiers* were hired and worked in the Moulin-Quignon pit until five that evening. Under close scrutiny a clean, vertical face of gravel was cut and from it five classic flint axes were found, sufficiently similar in shape to those claimed in Paris as forgeries. This was enough for Prestwich, the unwilling participant, to change his mind: 'No fraud had been practised', he declared, although 'my friends Dr Falconer and Mr Busk held a different opinion.'[136] If, he went on, the flint tools were genuine, then so too was the human jaw.

Prestwich had split the English camp. Falconer wrote daily to Grace, marking the letters *Private* and unburdening himself as the case unravelled. He and Busk stood firm, but none of the French wavered. The commission was now split five to three, with Carpenter back in England falling in with Falconer and Busk. Milne-Edwards summed it up in his report submitted to the Académie des Sciences on 18 May. The language remained fraternal but firm:

All the members of this gathering of friends embraced the same opinion. Dismissing any idea of fraud, they recognised, in the most candid way, that there no longer seemed to be any reason to call into question the authenticity of the discovery made by M. Boucher de Perthes of a human jaw in the lower part of the large deposit of gravel, clay and pebbles of the Moulin-Quignon quarry.[137]

Falconer was bloodied by the defence of 'the honour of an amiable old gentleman and the susceptibility of a fiery nation'[138] but unbowed, writing to Grace on 14 May: 'Therefore, my good Deductive, do not be cast down although your Uncle has been concerned as a loser in a 'Cause Célèbre.'[139] Hooker wrote caustically to Darwin on 24 May:

What a mess Falconer, Busk, Carpenter and Prestwich have made of it! . . . I cannot but think that Prestwich's position is most awkward—he had just claimed from Lyell the lion's share of authority in such matters and forthwith breaks down on a practical question.[140]

Prestwich's clenching authority, so admired by Falconer, no longer clenched. The two controversies of 1863 concerning the authority to speak on matters of human antiquity now swirled together, and the biggest casualty was Prestwich, dismissed by Hooker as someone 'ready to believe anything!' Joseph did not help himself, when in October he contradicted his June statement to the Geological Society by adding a last-minute note. He had changed his mind for a third time in eight months and now believed 'we were mistaken in concluding on that occasion that no fraud had been practised'.[141] The paper now betrayed muddled thinking, starting with one conclusion and ending with its opposite. And to confuse matters further, this change of heart applied only to the flint tools and not to the jaw. And who perpetrated the fraud? He implied in his latest volte-face that the *terrassiers* were cleverer than the savants of the commission standing over them as the pit was investigated. They had all, including himself, been duped by the men whose testimony had been so valued four years before.

The final act of *l'affaire Moulin-Quignon* fell to Evans. He wrote to *The Athenaeum* on 6 June and again on 4 July to spell out his objections to the flints which had been found. He listed the criteria to distinguish forgeries from the genuine articles—fresh surface scratches, colour that washed off, deviations in shape from his template, and percussion marks on the flints that could only have been made with a metal hammer. Too many from Moulin-Quignon failed these tests.

He had the advantage of not being present and followed Joseph's lead, claiming that in the 'circumstances of great excitement, there may have been an error on the part of such able observers' as the commissioners who oversaw the work.[142] He visited the pit, accompanied by Lubbock and the indefatigable John Wickham Flower, and they found evidence of the fraud they suspected, principally colouring the artefacts that were found for them to look right. Flint Evans was dismissive. There remained 'not the slightest doubt on my mind that a fraud, and a most ingenious and successful one, has been practised by some of the Abbeville terrassiers.'[143]

In his second letter on 4 July he disagreed with his friend Joseph, who had written to *The Athenaeum* on 13 June, when he still believed the stone tools were genuine.[144] Now Evans delivered the knockout blow. He had hired Henry Keeping, who worked at Brixham Cave for Pengelly and whose knowledge of digging cave sediments and supervising workmen proved invaluable. Over four days from 3 June, Keeping dug with the *terrassiers*. Stone implements were found every day, but he knew he was being conned. In his report to Evans he concluded, 'I have every reason to believe that all the specimens I have brought from Moulin-Quignon were placed there on purpose for me to find.'[145] Evans invoked the last rites: 'I sincerely hope that the human jaw from Moulin-Quignon may from this time forward be consigned to oblivion. *Requiescat in pace!*'[146]

L'affaire Moulin-Quignon is the counterpoint to the proof of human antiquity by the same time revolutionaries four years previously. Its claims were thoroughly investigated by geologists, archaeologists, anatomists, and palaeontologists. They may have been after the truth, but their greatest concern was that this controversy should not invalidate the hard-won time revolution of 1859. The well-known deniers of human antiquity were quick to seize on the confusion. Jacques's bitter opponent, Élie de Beaumont, denounced Moulin-Quignon at the Académie des Sciences as a modern peat bed more likely to produce Roman coins than antediluvian axes.[147] Quatrefages stood by the commission's findings but the critique went public.[148]

A fraternal stalemate had been reached, with the principals agreeing to disagree. After the commission had visited and pronounced upon the jaw, human remains continued to be found at Moulin-Quignon, and Boucher de

Perthes gladly acquired them.[149] This led John Evans to remark wryly in 1865 that ever since little or no gravel had been dug at Moulin-Quignon 'human bones have cropped up like mushrooms on its banks', including a complete skull.[150] All that survives of this large collection is twenty-eight specimens still in their original round cardboard boxes in the Muséum national d'Histoire naturelle in Paris. These include the jaw, and in 2016 the entire collection was subjected to its first forensic analysis since 1863.[151] Age is now determined by science-based methods such as radiocarbon dating, which uses tiny samples taken from the bones themselves. Even so, the jaw was regarded as too significant an historical object to extract a sample for dating. But two other bones, a fragment of skull and a tibia were dated. The results vindicated Falconer, Busk, and Tomes. The skull fragment was medieval, no more than 750 years old, while the tibia was even more recent; the person was alive sometime after 1664 AD in the reign of Louis XIV.[152] The forensic team concluded they had all come from a nearby cemetery, possibly one of Abbeville's plague pits used during the Black Death in 1346.[153] On a brighter note, among the stone tools in the same collection the archaeologists in the team found several genuine specimens, and these now ensure that Moulin-Quignon takes its place as an important early site in the Somme valley, visited 600,000 years ago.[154]

Belief in the jaw ebbed away as its supporters in France passed on— Boucher de Perthes in 1868, Lartet 1871, Bourgeois 1878, Delesse 1881, Desnoyer 1887, and finally Quatrefages in 1892. By the turn of the century Moulin-Quignon was forgotten, replaced by uncontested abundant human remains from the caves of France arising from Lartet and Christy's pioneering work in the Dordogne.[155]

But was Evans correct in fingering the *terrassiers* as Boucher de Perthes's mythical *mystificateurs* [hoaxers]?[156] They had the opportunity and they had the financial motive. But the planting of the jaw and the choice of specimen seem more scientifically knowledgeable, a point made by Falconer to *The Times*. Alfred Tylor hinted that, among the experts, Buteux and an English visitor with a geological interest, Nicholas Brady, found implements in the black seam where Jacques had found little: 'It seems strange', he observed ironically, 'that these gentlemen should have been so fortunate.'[157] But this is no smoking gun pointing at the culprits. And even if Élie de Beaumont was somehow, and it seems highly improbable, in cahoots with the *terrassiers*, then he singularly failed to capitalize on the controversy and use it to discredit Boucher de Perthes and the time revolution.

The problem was that nobody approached Moulin-Quignon with an open mind. There was a distinct lack of applying the Royal Society's motto 'Nullius in verba' ('Take nobody's word for it') to the question of authenticity. These were not, as was the case with the rediscovery of John Frere's Hoxne letter in 1859, either antiquaries or geologists unfettered by theory. They now believed in the great antiquity of humans, and this belief coloured judgements on both sides. Established positions had to be defended rather than demonstrated for the first time. With blinkers in place it proved possible for hoaxers, with some well-planted axes, to hoodwink members of the commission who desperately wanted to believe in Moulin-Quignon for the scientific glory of France as much as their commitment to the time revolution. But not Henry Keeping and certainly not Evans, Lubbock, and Flower, who came with a battery of criteria. Neither camp was ever likely to change its mind. Only Prestwich, whose equivocation so exasperated Falconer, but who, nonetheless, was soon forgiven. But such flip-flopping did have an unfortunate outcome for Joseph's reputation. It granted control of the time revolution to Lyell through his book, which had given such offence. Today, Lyell is widely remembered and commemorated; Prestwich is not.[158]

On one point alone was there agreement: Jacques, the 'Patriarch of Primeval Archaeology',[159] though a *gobemouche*—overtrusting and overemotional—was blameless of fraud.

Notes

1. Van Riper 1993: 153.
2. See Chapter 6.
3. Delanoue made his frustrations known at a meeting of the *Société d'anthropologie* in Paris (Hurel and Coye 2016: 317).
4. Prestwich, Letter to his sister Catherine, 31 March 1860 (G. A. Prestwich 1899: 147–8). Catherine Prestwich (1815–88) married Robert Thurburn in 1839.
5. G. A. Prestwich 1899: 263.
6. McNabb 2012.
7. The report of the Water Commissioners came out in 1869.
8. The report of the Coal Commissioners came out in 1871.
9. Prestwich 1874 see also G. A. Prestwich 1881[1901] for her informed and witty account of past attempts to bridge the Channel.
10. The Kent's Cavern committee reported every year to the British Association for the Advancement of Science on new discoveries. The other inaugural members of the committee were John Phillips, president of the Geological Society, Edward

Vivian, geologist. Pengelly was the designated reporter for the committee and the most experienced of the excavators due to his involvement with the Brixham excavations.

11. These new investigations built on the earlier, but overruled, findings of the Rev. John MacEnery (see Chapter 3). His work was lauded by Prestwich in his presentation to the Royal Society in May 1859.

12. Evans 1866a: 173. The purpose of the working model was to show how the position of the polar axis could be changed with small increments in mass as represented by a series of adjustable screws with heavy heads. A change in this axis was, Evans believed, a factor in pushing climate into an ice age.

13. Hugh Falconer, Letter to Evans, 15 January 1863, Ashmolean Museum, *Evans Collection*.

14. Evans 1865. Boylan (2008: 59–63) provides the background to Evans's investigation. *Archaeopteryx* had to wait until 1954 for its 'definitive' study by Sir Gavin De Beer. The wait was worth it because Evan's interpretation was preferred over Owen's.

15. *Darwin Correspondence*, Darwin to Henslow, 26 October 1860.

16. *The Times*, 22 March 1850, p. 5.

17. A flashpoint for the Civil War was John Brown's raid and death at Harpers Ferry in October 1859.

18. The colonial wars of the 1860s boosted the production of cartridge paper in Dickinson's paper mills. Indeed, it might even have been some of Evans's cartridges coated in animal fat that, when bitten into, caused the flashpoint for the Indian Rebellion in 1857. Lubbock's father-in-law Augustus Lane-Fox wrote the manual on musketry for the British Army in 1855. It included a template for making paper cartridges (1855: 25–9) with this instruction: 'when completed, the base of the cartridge must be dipped up to the shoulder of the bullet in a pot of grease, consisting of six parts tallow to one of bee's wax' (1855: 29). The manual on musketry was published anonymously.

19. https://en.wikipedia.org/wiki/List_of_wars:_1800%E2%80%931899#1860%E2% 80%931869, accessed 1 August 2020.

20. Mark Carney, Governor of the Bank of England, illustrated his speech on 5 December 2016, *The Spectre of Monetarism* (https://www.bis.org/review/ r161207d.pdf, accessed 1 August 2020), with the run on this bank. See https:// www.bankofengland.co.uk/-/media/boe/files/quarterly-bulletin/2016/the-demise-of-overend-gurney, accessed 1 August 2020.

21. Wages fell drastically in 1860 and took a decade to recover; see https://www. theguardian.com/business/2016/dec/06/landscape-exhaustion-moral-decay-1860s-lost-decade-mark-carney, accessed 1 August 2020.

22. For John Evans's leading role in the repeal of the duty on paper in 1861, see Evans 1955: 103–12; Cannadine 2018: Kindle p. 5461.

23. Evans 1943: 121.

24. *Darwin Correspondence*, Darwin to Prestwich, 12 March 1860. Prestwich's reply, if he made one, has not survived.

25. Darwin wrote two letters to Henslow with passing reference to his mentor's letters in *The Athenaeum*. *Darwin Correspondence*, Darwin to Henslow, 26 October and 10 November 1860.

26. Henslow, 'Flints in the Drift', Letter to *The Athenaeum*, 20 October 1860, p. 516 and 3 November pp. 592–3.

27. Henslow, *The Athenaeum*, 20 October 1860, p. 516: 'The facts I have witnessed do not *of necessity* support the hypothesis of a pre-historic antiquity for these works of man.'

28. Rogers has been inducted into the Creation Science Hall of Fame partly for his geological defence of the biblical Flood and his signing of the 1865 Declaration affirming the scientific integrity of Scripture; see https://creationsciencehalloffame. org/piece/henry-darwin-rogers, accessed 1 August 2020 Darwin, however, in a letter to Hooker, *Darwin Correspondence*, 3 March 1860, believed Rogers was partly convinced of his evolutionary theory.

29. Rogers 1860: 438–9.

30. Rogers 1860: 439:

> In conclusion, then, of the whole inquiry, condensing into one expression my answer to the general question, Whether a remote prehistoric antiquity for the human race has been established from the recent discovery of specimens of man's handiwork in the so-called Diluvium, I maintain it is not proven, by no means asserting that it can be disproved, but insisting simply that it remains—*Not proven*'.

31. This magazine for the highbrow was published quarterly between 1854 and 1865.

32. Lubbock 1862: 247.

33. Lubbock 1862: 239, quoting Rogers 1860.

34. Evans 1863: 65–6.

35. See Chapter 1.

36. In the later *Transactions* paper, 'Drift' became ('Drift*') and he announced in a footnote that this term should be dropped as imprecise. He doesn't tell us what should replace it, although in the earlier *Proceedings* version he does use Lyell's term Pleistocene, which is the term used today.

37. Loess, also known as brickearth, is another superficial but in this case windblown deposit.

38. Lyell, Letter to the Royal Society, 8 June 1863, Royal Society Library, Referees Reports, RR4, p. 208.

39. James Geikie (1894) and others realized these fine sediments owed their origin to wind rather than water.

40. Prestwich identified the action of ice in river gravels at Joinville in the Somme Valley, where he found a massive sandstone boulder that was too large to be

moved by a river. Lubbock agreed with his interpretation that it had been moved there by river ice.

41. Bridgland 2014: 51. The skull was described by the palaeontologist Richard Owen in 1856.

42. Sir Archibald Geikie provided an overview of Prestwich's Pleistocene geology (G. A. Prestwich 1899: 402–21).

43. Antoine et al. (2011) recount the changing interpretations of river terraces by French and English geologists.

44. Prestwich 1864a: 286.

45. Prestwich 1864a: 286.

46. Lubbock 1862: 258.

47. Prestwich 1862: 49.

48. Tyndall 1871: ch. 13; Owen 2013: 36–7.

49. Tyndall 1871: 144.

50. Tyndall 1871: 150.

51. 'Christian Manliness' was Kingsley's preferred term, but 'Muscular Christianity' has stuck (Conlin 2014: Kindle p. 2605).

52. They set out to discover whether the rate of combustion of a candle varies with the density of the atmosphere in which it is burnt, a question which was answered in the negative during their night atop Europe's highest mountain; see https://en.m.wikipedia.org/wiki/Edward_Frankland, accessed 1 August 2020.

53. G. A. Prestwich 1899: 116 and 139.

54. Prestwich 1864a: 305.

55. See Patton (2016) for Lubbock's deep-seated but rather vague Anglican beliefs, and Evans (1943) for her father's arm's-length religion.

56. Milne 1901: 5–6.

57. Milne 1901: 7.

58. This important sentence is missing from the letter reproduced in Milne (1901: 7) but is in the version printed in Murchison (1868: 52).

59. No. 21 Park Crescent by Regent's Park. It is another terrace designed by John Nash and, like the nearby Kent Terrace where the Prestwiches lived, has always been prime London real estate. The house was owned by Charles Falconer, Hugh's brother.

60. G. A. Prestwich 1899: 21.

61. Tabrum 1910.

62. Tabrum 1910: 26.

63. Evans 1964: 37.

64. Clark 2014: 73. The quotation is from Lubbock 1887: 167.

65. Trautmann 1992: 383.

66. Lyell, Letter to Hooker, March 1863 (Clark 2014: 73).

67. Trautmann 1992: 383.

68. Conlin 2014: Kindle p. 2871. See Chapter 1 for a description of the big ditch.

69. Bishop Wilberforce's long account of creation rather than change ended with his question whether the apes were on Huxley's grandfather's or grandmother's side of the family. Huxley responded:

> Would I rather have a miserable ape for a grandfather or a man highly endowed by nature and possessed of great means of influence and yet who employs these faculties and that influence for the mere purpose of introducing ridicule into a grave scientific discussion, I unhesitatingly affirm my preference for the ape.   (Desmond 1998: 279)

Lubbock and Hooker, close allies of Huxley in Darwin's camp, spoke after him, and, as Barton points out (2018: Kindle p. 3627), the outcome was not a defeat by Huxley of the bishop, but a combination of Huxley, Hooker, and Lubbock.

70. Owen believed in continuous creation and transitional fossils (Desmond 1998: 274). But humans were a thing apart and needed some additional spark. As Desmond (1982: 36) describes him, 'Owen comes across as a psychologically complex character struggling with immensities—driven by his hatred of ape ancestry.' This led him to look for a non-transmutation theory to counter Darwin's evolution of species through gradual change.

71. These are covered in Desmond (1998).

72. *Monkeyana* was written by Sir Philip Egerton, former president of the Geological Society (Van Riper 1993: 173 n. 52). The *Monkeyana* cartoon is seen by Holterhoff (2016: 217) as a response to the outbreak of the American Civil War a month before at Fort Sumter. If so, it was a satiric challenge to the master–slave relationship between the human races, the staff suggesting action rather than supplication.

73. Pengelly, Letter to his son, 11 May 1861:

> I reached town just in time for dinner, after which we all went to a large party at the Bishop of London's. I met Murchison there, who told me that all the geologists connected with the flint instrument question are to be immortalized in 'Punch' next week. Your dad in the number.
>
> (Pengelly 1897: 104)

74. Dickens 1865: 297.

75. Bennett 1872.

76. Lubbock 1862: 250. This was a belated response to Richard Cull's letter to the *Athenaeum*, 25 June 1859, where the point had been raised.

77. Lubbock 1862: 267. He argued that cold northern climates were unsuited to early humans and that a productive search would be to look in warmer, southern climates. While Darwin is correctly credited with directing attention to Africa as an evolutionary centre for humans, Lubbock could lay claim to that prediction also.

78. An upper jaw was found in Kent's Cavern in 1927 and may date to 40,000 years ago.

79. Oakley 1964.

80. Human remains had also been found elsewhere in France during the excavation of caves at Arcy-sur-Cure, Yonne (1859 and 1863), Arudy in the Basses-Pyrénées (1863), and earlier in Wales at Goat's Hole, Paviland (1822–3) (Oakley et al. 1971).
81. Oakley, Campbell, and Molleson 1971.
82. Pitts and Roberts 1997. This ancient shoreline in front of a collapsing chalk cliff was dug during the 1980s and 1990s.
83. The type fossil of *Homo heidelbergensis* (Oakley et al. 1971).
84. *Darwin Correspondence*, Darwin to Huxley, 18 February 1863. He had received a copy in advance of publication.
85. Lyell 1863: 506. In these pages he mentions evolutionary succession (transmutation) and is largely paraphrasing the views of the American naturalist Asa Gray, regarded by Darwin as an ally.
86. Lyell 1864: 74. Van Riper (1993: 153) seizes on this observation as indicating Lyell's preferred age of 100,000 years for the Pleistocene.
87. Lyell 1863: 202. He was comparing the Natchez loam with Mastodon bones to the Mammoths found in the Somme and Thames alluvium.
88. Lubbock 1865: 216.
89. Huxley 1863: 114–20.
90. Boylan 2008: 61–2.
91. Evans 1865: 421. His publication was delayed by two years partly because of Owen's claims on first publication of the fossil. This took place in 1863 in the *Transactions* of the Royal Society. Owen dismissed Evans's claim that the jaw and teeth belonged to *Archaeopteryx*, identifying them instead as those of a fish. This led Evans to undertake further detailed studies. He submitted his paper in 1864, and it was published the next year (Boylan 2008: 61).
92. Busk 1861.
93. Schaaffhausen (1858), translated by George Busk in 1861. The skull was brought to Schaaffhausen's attention by J. C. Fuhlrott in 1857. It was named as a separate species *Homo neanderthalensis* by the anatomist William King in 1863 (Oakley 1964: 110).
94. Oakley 1964: 110.
95. Huxley in Lyell 1863: Fig 4, p. 83.
96. Huxley in Lyell 1863: Fig.6, p. 88.
97. Huxley 1863: 159.
98. Huxley 1863: 159.
99. Lubbock's review of Lyell appeared anonymously in *The Natural History Review*, April 1863, where he wrote about Lyell's views on *The Origin*: 'We are, however, unable to discover that Sir Charles anywhere expresses his own opinion' (p. 213). This was a book of which 'even the Bishop of Oxford might approve' (Patton 2016: Kindle p. 238).

100. Falconer, Letter to *The Athenaeum*, 4 April 1863, p. 460. Lubbock agreed, writing to Darwin on 7 April, 'He [Lyell] certainly has not quite done justice to others and particularly to Prestwich' (*Darwin Correspondence*).

101. Lubbock's interest in antiquities started in 1860, so he could also have been open to the charge of jumping on the Evans-Prestwich bandwagon that started rolling in 1859, except he did not seek to control the time revolution in the way that Lyell did (Owen 2013).

102. Lyell, Letter to *The Athenaeum*, 18 April 1863, p. 524, letter written on 15 April.

103. Lyell, Letter to *The Athenaeum*, 18 April 1863, p. 524.

104. Lyell, Letter to *The Athenaeum*, 18 April 1863, p. 525.

105. Prestwich, Letter to *The Athenaeum*, 25 April 1863, p. 555.

106. *Darwin Correspondence*, Darwin to Hooker, 17 April 1863.

107. *Darwin Correspondence*, Hooker to Darwin, 29 March 1864.

108. Falconer, Letter to McCall, 8 February 1863 (Boylan 1979: 179).

109. Falconer, Letter to *The Athenaeum*, 2 May 1863, p. 586.

110. Falconer, Letter to *The Athenaeum*, 2 May 1863, p. 586.

111. Joseph attacked him in *The Athenaeum* in late April. Lyell overturned Hopkin's negative report in June.

112. Falconer wrote a long account in 1863 of the matter, *Primeval Man, and his Contemporaries*, which pointedly relegates Lyell's *Antiquity of Man* to a footnote on page 588. The paper was published posthumously in 1868. Grace Prestwich touches on the bitter accusations (1899: 176–7) with the tone of someone who had had an earful at the time from Uncle Hugh. Long after Falconer had died, Lyell recast in the fourth edition of *Antiquity of Man* in 1873 the sections on Brixham. Grace at least was satisfied with his climbdown: 'Prestwich and Falconer had been the pioneers in the inquiry' (G. A. Prestwich 1899: 177).

113. The Moulin-Quignon story is told effectively and in great detail by Boylan (1979); Cohen and Hublin (1989); Hurel and Coye (2016); Hurel et al. (2016).

114. Hurel and Coye 2016: 317. For a contemporary account, see Alfred Tylor (1863) and Boucher de Perthes (1864a).

115. Tylor 1863.

116. Tylor 1863: 168.

117. Boucher de Perthes, Letter to Falconer, 26 April 1863 (Boucher de Perthes 1864a: 614).

118. *Darwin Correspondence*, Boucher de Perthes to Darwin, 23 June 1863.

119. Falconer changed his mind quickly on his return to London but recorded his initial thoughts in a letter to *The Times*, 'The Human Jaw of Abbeville', 21 May 1863, p. 13.

120. Boylan 1979: 180. Falconer, Letter to Prestwich, 16 April 1863. Falconer Museum, Forres.

121. Carpenter, 'Discovery at Abbeville'. Letter to *The Athenaeum*, 18 April 1863: 523. The letter was sent from Abbeville on 14 April.
122. Falconer's letter was seen by the editor as both important and good for circulation, and space was made for it by dropping the daily Law Report (Boylan 1979: 181).
123. Falconer, Letter to *The Times*, 25 April 1863.
124. The letter to *The Times* misprints his name as Somes.
125. Falconer, 'The Reputed Fossil Man of Abbeville', *The Times*, 25 April 1863, p. 14.
126. Boucher de Perthes (1864a: 613–7, translation in Boylan 1979: 182). The seance took place in Paris and the letter was dated 26 April 1863, reaching Falconer the next day.
127. Falconer, Letter to McCall, 28 April 1863 (Boylan 1979: 183).
128. Falconer, Letter to *The Times*, 'The Human Jaw of Abbeville', 21 May 1863, p. 13.
129. Falconer 1868: 596.
130. Boylan 1979: 186. Falconer approved, writing to Grace on 10 May that he was 'half-English'.
131. See Chapter 2.
132. See Chapter 4.
133. Boylan 1979: 185.
134. Boylan 1979: 190.
135. As Boylan says, this was preposterous (1979: 189–90). Jacques had been looking for these remains for over thirty years, and Quatrefage's explanation pointed the finger at him, whereas Falconer had only blamed the *terrassiers* for the suspected fraud.
136. Prestwich 1863: 499. Paper read to the Geological Society on 3 June.
137. Boylan 1979: 192. The full text and much supporting evidence can be found in Boucher de Perthes (1864a: 179–93), the third and final volume of his *Antiquités celtiques et antédiluviennes*, which deals at length with Moulin-Quignon.
138. Falconer, Letter to McCall, 15 May 1863 (Boylan 1979: 192).
139. Boylan 1979: 191.
140. *Darwin Correspondence*, Hooker to Darwin, 24 May 1863.
141. Prestwich 1863: 505.
142. Evans, 'The Abbeville Human Jaw', Letter to *The Athenaeum*, 6 June 1863, p. 747.
143. Evans, 'The Abbeville Human Jaw', Letter to *The Athenaeum*, 6 June 1863, p. 748.
144. Prestwich, 'The Human Jaw of Abbeville', Letter to *The Athenaeum*, 13 June 1863, pp. 779–80.

145. Evans, 'The Human Remains at Abbeville', Letter to *The Athenaeum*, 4 July 1863, pp. 19–20. It was Keeping's evidence that led to Prestwich's volte-face later in the year.

146. Evans, 'The Human Remains at Abbeville', Letter to *The Athenaeum*, 2 July 1863, p. 20.

147. Falconer, Letter to Evans, 29 May 1863 (Boylan 1979: 196).

148. The full intervention is presented by Hurel and Coye (2016: 331).

149. Boucher de Perthes 1864a: 215–52. Hurel and Coye report that over a hundred human remains were found between 19 April and 16 July 1864 (2016: 336).

150. Evans 1866b: 363.

151. Vialet et al. 2016.

152. Vialet et al. 2016: 416–17.

153. Hurel et al. 2016: 309.

154. Bahain et al. 2016; Moncel et al. 2016.

155. Lartet and Christy 1865.

156. *Darwin Correspondence*, Boucher de Perthes to Darwin, 23 June 1863, where he dismisses the idea of hoaxers.

157. Tylor 1863: 167.

158. Today this would be a battle over intellectual property, and the lawyers would be involved. As a further sign of their relative status, Lyell has a London blue plaque, even though the building where he lived between 1854 and his death in 1875 is no more, while Prestwich's house at Kent Terrace, which survives him, does not (see Chapter 7).

159. Tylor 1863: 168.

6

Acceptance

The Decade Closes 1864–72

The wind had left the sails of the Moulin-Quignon. It had also deserted Joseph Prestwich as he havered over the authenticity of its bones and stones. But the time revolution blew on without him. John Evans collected more flint implements, the ballast of deep history, while John Lubbock manoeuvred the revolution into one of the swirling currents driven by the demonstration of unlimited geological time. On Lubbock's watch the possibilities of the time revolution now shifted to the greatest moral challenge facing nineteenth-century scientists: the deep history of race.

The Cavemen of the South

In March 1864, five years after the time revolution which he helped start, Hugh Falconer changed tack. He stopped savaging Lyell and left Boucher de Perthes alone. Cordial relations resumed with his good friend Édouard Lartet, who had been on the opposite side of the Moulin-Quignon Commission. So much so that he sent a letter to *The Times*,[1] drawing attention to the astonishing discoveries made in the last five months of the previous year by Henry Christy and Lartet in the Vézère Valley of the Dordogne region of France.[2] In its meandering course the river had scooped out the limestone cliffs which line it to form deep, overhanging rock shelters. Antiquaries like Lartet and Christy could dig here without getting wet. In 1863 when their excavations began, these rock shelters, *abris*, were used by farmers to store produce and overwinter livestock, and many served as *caves* for the wine of the region. What they found reflected the richness of the land. On the right bank of the river they investigated *abris* at Laugerie Haute, Laugerie Basse, and La Madeleine which could not have been more different to the bleak gravel pits of the Somme. The implements were smaller, still unpolished, but finely made long, thin knife blades of flint belonging to the Stone Period.[3] There were even some genuine ancient human remains from

La Madeleine located at a strategic ford across the great loop in the river.[4] The limestone *abris* preserved animal bone especially well, and Lartet used their abundance to cut the chronological cake of the Stone Period. He started with an age of extinct animals—mammoths, cave bears, and woolly rhinos— into which Prestwich's high terrace of the Somme also fitted. Then came the Reindeer Age, named after the many thousands of bones of this species they dug from *abris* such as La Madeleine and Laugerie Haute. And finally came the age of domestic animals. It was the Reindeer Age that merited a letter to *The Times*, for here, and elsewhere at the Bruniquel cave in the Averyon Valley, Lartet and Christy had found the first ice age art. This included engravings on stone slabs and carved bones and reindeer antlers depicting the animals whose bones lay alongside them. And in May 1864, after *The Times* letter, he went with Lartet and Christy to La Madeleine, arriving just after the workmen had discovered five broken pieces of a thin slab of mammoth ivory on which engraved lines were visible (Figure 6.1).[5] The pieces could be refitted, and Lartet handed these to Falconer. The authority on ancient elephants instantly saw the head of a mammoth and its woolly coat. Uncle Hugh was holding an engraving by an eyewitness of an extinct pachyderm.

And there was more: harpoon heads made of antler, fish bones, sewing needles of bone and ivory, and fireplaces that turned rock shelters into homes. Christy, second only to Evans as an authority on flints, was well travelled in Mexico and North America. And from these experiences he drew comparisons with the artefacts and settlements of their indigenous peoples. He was particularly struck by the resemblance in animals, technology, and habits of the Eskimo (Inuit) and the well-clad cave dwellers of France during the cold Reindeer Period.[6]

The discoveries in the Dordogne offered another vision of Primeval Man. Christy summed it up in the last sentence of his paper in the *Transactions of the Ethnological Society of London*:[7] 'We are bound to confess that, so far, nothing in the investigation of the works of uncivilized or primitive man, either of ancient or modern times, appears to necessitate a change in the old cherished idea of the Unity of the Human Race.'[8] This idea was *the* ethnological problem, championed by Prince Albert at the start of the 1850s, and where the task was to demonstrate the unity of the human species.[9]

Not so according to the Mosaic account of human origins shoehorned into 6,000 years of recorded biblical history. The history of the human races began physically with Noah's sons and culturally with the Tower of Babel when a shared language was overthrown (Table 4.1).[10] The assorted

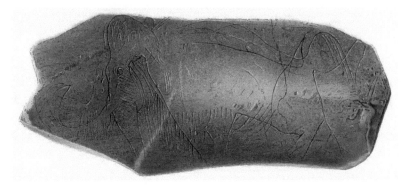

Figure 6.1 The engraving of a mammoth on a slab of mammoth ivory found in 1864 at the La Madeleine rock shelter in the Dordogne. It was seen by Lartet and Falconer on the day of its discovery. The head and tusks of the mammoth are facing left.
Source: Lartet 1865–75: Plate XXVIII.

nineteenth-century races—African, Asian, Australian, American, and European—were, according to Genesis, original, like the roots of closely planted, but different trees. This was the view of the polygenists, who formed the majority on matters of race, and when the timescale for human antiquity was short, the idea was plausible. But extend the timescale to geological proportions and the opposite views of monogenists like Christy, that the geographical races had differentiated from a single common source, became scientifically inevitable.

Henry Christy's conclusion was based on the hard evidence of Stone Age artefacts. He proposed three periods; Drift, Cave, and Surface.[11] The Somme and Hoxne fitted into the first and were the oldest geologically. Then came his work with Lartet along the Vézère and to which the Engis caverns in Belgium could also be added and some of the Welsh caves on the Gower Peninsula, most notably at Goat's Hole, Paviland.[12] These possessed a complex range of bone, ivory, and antler tools, as well as the new forms of struck flints found in the Vézère *abris*. Finally, surface finds, the youngest of Christy's three periods, started with the coastal Danish shell middens, which lacked reindeer bones and flourished among the Swiss Lake villages with pottery, polished axes, and crops and domestic animals.[13]

Here was Spencerian progress from simple to complex. For time revolutionaries like Falconer, Christy, and Lartet the progression was clear. It went from coarse and heavy handheld implements such as those in northern

France and southern England to the lightweight tools, hafted into spears and possibly arrows, of southern France. Furthermore, the superior tool kit was embellished by art. And if there was art, were they not dealing with people closely linked to ourselves with language and possibly a religious sense, as indicated by aesthetics and bodies carefully buried in special places in the Vézère's *abris*?

What Did They Look Like?

Humanity might have progressed, anatomically and technologically, but this did not change attitudes to race which brutalized millions of lives throughout the nineteenth century. New sciences, such as anthropology and prehistoric archaeology, were called on to normalize these attitudes and justify unequal power relations between the colonizer and the colonized, the slave and the master. The wider context to these scientific debates in the 1860s was the American Civil War, which by 1864 was in its third full year of carnage and blockade. There was famine in Lancashire, as cotton supplies were interrupted, and the British mood was unfavourably disposed to the Northern cause.[14] Many would not have been laughing at *Punch*'s sandwich-board gorilla in the satirical *Monkeyana* but at the idea of emancipating the Southern slaves.

None more so than Dr James Hunt (1833–69), who took the opportunity in his introductory address to the Anthropological Society of London, which he founded in 1863,[15] to write, 'I would therefore express a hope that the objects of the Society will never be prostituted to such an object as the support of the slave trade, with all its abuses; but ... ', and here came the justification used ever since by racists and white supremacists, 'at the same time we must not shirk from the candid avowal of what we believe to be the real place in nature, or in society, of the African or any other race.'[16] Anything else would be the Victorian version of political correctness or, as Hunt called it, 'rights-of-man-mania'.[17] What he really meant was spelt out in a paper to the same learned society in November 1863, with its sneering title, *On the Negro's Place in Nature*, which was clearly aimed at Huxley, his bête noire.[18] Hunt's conclusions, based on anatomical distortions unforgiveable in a medical man,[19] led him to state that Africans were a distinct species from Europeans in which 'the analogies are far more numerous between the Negro and apes than between the European and apes'.[20] He concluded they were intellectually inferior to the European, who could humanize and civilize them, even though they were unsuited to European civilization. And if his racist rant was not bad enough, it was ringingly endorsed in the following forty pages of comments by other members of his scientific society.

Even by Victorian standards, where racial stereotypes were accepted without argument and their vocabulary shocks, Hunt was a grotesque who might be dismissed as the pantomime villain of anthropology if his legacy has not proved so enduring. His Anthropological Society, a breakaway group from the older Ethnological Society, had a Confederate agent sitting on its council and a slush fund traced back to Richmond, Virginia.[21] While they paid lip service to abhorrence of the slave trade, they set out to provide scientific cover for the continuing institution of slavery.

What of the opposition with its single origin for humanity subsequently differentiated by natural selection? Alfred Wallace ventured into Hunt's unsavoury society to give, in 1864, a paper on *The Origin of Human Races and the Antiquity of Man Deduced from the Theory of Natural Selection*. Wallace had lived among peoples from many races when collecting floral and faunal specimens for the museums of Europe. He had seen natural selection in action, where greater numbers of Europeans and their individual longevity and capacity for increase explained, for him, their success in the nineteenth century's struggle for global existence. Looking back into remote antiquity, Wallace argued that humankind started as a homogeneous race without the faculty for speech and living in a tropical region. Subsequent differentiation into the geographical races of 1864 was due in his opinion to natural selection working on skin colour, stature, and other physical attributes to produce the observable differences, 'modified in accordance with local conditions'.[22] It was only later, he went on, as people migrated away from warmer climates,[23] that the evolution of the mind occurred, and this allowed humans to adapt and diversify culturally and technologically so that physical evolution was no longer the prime means of survival. This was for Wallace 'the true grandeur and dignity of man', where, 'from the moment when the first skin was used as a covering, when the first rude spear was formed to assist in the chase...a being had arisen who was no longer necessarily subject to change from the changing universe'.[24]

However, the Wallace-Darwin marriage of unlimited time and natural selection went down like a lead balloon before Hunt's time-revolution-hating anthropologists. They battered Wallace with questions and tried to trip him with non sequiturs as they paraded their prejudices on the meaning of human variation. He got progressively irritated as their barrage pounded against his scientific argument 'that man without speech is not man',[25] and so subject to natural selection like any other creature.

Stone Ages and Weeds 1864–5

The nurseries of Nash House and the Lubbocks' large new home, Lamas in Chislehurst, Kent were full of children at Christmas 1863. Harriet and Alice Evans were 6 and 7 and their two brothers Philip, and Lewis 9 and 10, while Arthur, now 13, had moved downstairs. Nelly Lubbock's children were tightly spaced; baby Gertrude born that year, then Norman, Constance, John, and the oldest, Amy Harriett, who was 6.

What stories might the children of time revolutionaries read and be read to? The obvious choice in 1863 was Lubbock's friend, the Rev. Charles Kingsley's *Water Babies*, serialized since 1862 in *Macmillan's Magazine* and now issued as a book.[26] *Water Babies* is no longer read unabridged because its adventures of a chimney sweep, Tom, and his friend Ellie contain too many ugly stereotypes and racist opinions[27] as they climb the evolutionary ladder from amoeba to human.[28] Kingsley, the original muscular Christian, had been treated in 1859 by James Hunt for his stammer.[29] Some of Hunt's opinions he shared, but they would have fallen out over Kingsley's enthusiasm for the time revolution[30] and Darwin's natural selection.

Water Babies is a very Samuel Smiles, *Self-Help* tale, replete with uplifting aphorisms for young minds such as Mother Carey declaring, 'Know silly child...that any one can make things, if they will take time and trouble enough: but it is not everyone who, like me, can make things make themselves.'[31] And these young minds lapped up the cautionary tale of the Doasyoulikes, who had left the land of Hardwork and suffered the fate of those who only did what pleased them. They lived in the land of Readymade at the foot of the Happy-go-lucky Mountain, living off flapdoodle, which they didn't even have to grow. The fairy who shows the children the waterproof book of the Doasyoulikes, full of photographs invented '13,598,000 years before anybody was born',[32] then turns over the pages at 500-year intervals to expose their slow decline, until all that was left was one tremendous old fellow, 7 feet tall. He was shot, thumping his chest, the book tells us, by Paul du Chaillu (1831–1903), famous for discovering the gorilla in the wild. The old fellow had tried to stop the zoologist's bullet by appealing to him, 'Am I not a man and a brother?' but had forgotten how to speak.[33] In so doing the last Doasyoulike echoed Wallace, that a man without speech is not a man, the twisted joke of *Monkeyana's* gorilla. The moral was clear: behave badly and you suffer the worst evolutionary fate, degeneration, precisely what happens throughout the book to Tom the chimney sweep whenever he is bad.³⁴ Goodness, cleanliness, and hard work lead to evolutionary success. That was how muscular Christians climbed natural history's ladder, no doubt with Smiles's full approval.

While Lubbock's children were swimming through evolutionary fables upstairs at Lamas, downstairs in his study their father was writing about time. The result was his first bestseller *Pre-Historic Times: As Illustrated by Ancient Remains, and the Manners and Customs of Modern Savages* which appeared in the spring of 1865. It drew heavily on five of his published papers, including his study of the Somme terraces, and benefited from visits to the Dordogne, Somme, Denmark, and Switzerland to see the evidence at first hand. He began by introducing two new words to describe the remote past: Palaeolithic and Neolithic, the Old and New Stone Age. Palaeolithic covered the unpolished axes and flint implements of both the Somme and Dordogne. Neolithic took in the polished axes from the surface and the giant stone structures, megaliths, of which the most famous in Britain were Stonehenge and Avebury.

The structure of *Pre-Historic Times* was designed to convince any remaining sceptics of the antiquity of humans. Lubbock did this by devoting the first ten chapters to archaeology, where, in Evans's later phrase, he descended the stream of time tightly holding his reader's hand.³⁵ He started in the Bronze Age of the Northern Antiquaries and went down past the Swiss Lake villages, deeper to the Danish Shell Mounds, down again to the Pleistocene faunas and the caves of the Vézère until finally in chapter nine he reached Boucher de Perthes and the Somme. This part of the book concluded with an overview of geological age and human antiquity. In this way the reader passed from the familiar to the unexpected, from monuments like shell mounds and stone circles that could be visited on the surface down to La Madeleine's many levels until they bottomed out in the deep archaeology of the Fréville pit at Saint-Acheul. Here was an argument made by tracing a journey back through the layers of time.³⁶ He was asking the reader to accept a gradual continuity in deep history, rather than making one giant leap from the present into the abyss of time, as undertaken by Prestwich and Evans when they started the time revolution.

Pre-Historic Times was a travel book to remote time and distant places. Lubbock, the guide, took his readers on a tour of the principal archaeological sites and collections of Western Europe, accompanied by over 150 illustrations. In chapter eight he drew on other's accounts to range more widely into North America and in later editions incorporated new finds such as those made by the geologist Robert Bruce-Foote (1843–1912) from Pallavaram

(Chennai), India.[37] These unpolished axes, two of which Lubbock illustrated, fitted the ovate and pointed forms from the Somme described by Evans in his *Archaeologica* paper the year before.[38] The Old Stone Age now had a claim to be a global archaeology, an Imperial Palaeolithic for the deep history of his British readership.

But Lubbock had another aim in his book: to show that the 'most sanguine hopes for the future are justified by the whole experience of the past'.[39] A process which had been in motion for thousands of years was not going to stop suddenly.[40] That things would continue to improve for all mankind was John's understanding of how natural selection applied to both the body and, unlike Wallace, the mind.[41] In John's opinion any improve-ment to the mind led to an increase in human happiness. Education and science were the means to achieve this goal. This was the lesson the past taught him and which in the last three chapters of *Pre-Historic Times* he set out to show was the case in the modern world.

His starting point for the deep history of human improvement was the Palaeolithic of the Dordogne with its art, reindeer, and much-lived-in *abris*. Even so, this was a low state of civilization as judged by what came later.[42] He then did a tour of the world's peoples,[43] 'Modern Savages' in the judgemental terminology of 1865, who did not use metals. These included Pacific islanders, the Maori of New Zealand, South African Hottentots, the Veddahs of India, Arctic Inuit, and North and South Americans, all the way down to Tierra del Fuego (Figure 6.2). Many lived by fishing, gathering, and hunting. The polished axe, so common throughout the island societies of the Pacific, combined with domesticates such as breadfruit, pigs, and sweet potato, put these peoples into his Neolithic archaeological stage. Those at the uttermost ends of the earth in Tierra del Fuego, Australia, and Tasmania

Figure 6.2 Living prehistory illustrated in *Pre-Historic Times*. At the top is a bone harpoon from one of the ancient Danish shell mounds and underneath a longer but similar example which Darwin collected in Tierra del Fuego while on board the *Beagle*. He gave it to his young friend John Lubbock in 1864. The scale is two centimetres.
Source: Lubbock 1865.

lacked polished stone tools and domesticates, and their technology mirrored in his view the Palaeolithic of Europe. Tylor, the ethnologist, agreed, telling his readers that eyewitnesses had seen Tasmanians pick up suitable stones, knock off a few flakes, and use them to cut or notch wood.[44] And what's more, he continued 'there is a specimen corresponding exactly to this description in the Taunton Museum', where it can still be seen,[45] and 'an implement found in the [geological] Drift near Clermont [France] would seem to be much like this'.

It is difficult, however, to reconcile John's dismissal of many of the societies he touched on as 'miserable savages',[46] (he even drew up a league table of who was the least civilized),[47] with the positive message of improvement he derived from his historical and global survey. The reason for the light amongst the gloom was simple. Lubbock's 'Modern Savages' had not fallen from grace, degenerated like a bad Tom in his children's bedtime reading. Rather, they had not yet evolved to realize their potential. They were flies trapped in amber, and in one of his most quoted phrases, comparable to living fossils from the animal kingdom:

> If we wish clearly to understand the antiquities of Europe, we must compare them with the rude implements and weapons still, or until lately, used by savage races in other parts of the world. In fact, the Van Diemaner [Tasmanian] and South American [Fuegian] are to the antiquary, what the opossum and the sloth are to the geologist.[48]

Here, then, was an answer to that question 'What did they look like?' While Lubbock avoided drawing direct parallels between the Palaeolithic of the Somme and a living people, he nonetheless steered his readers, eager to find a face for the Palaeolithic, towards the modern world (Figure 6.3).

If Lubbock traced back to make his points about great antiquity and the unity of humans, another book which appeared that same year traced up. This was Edward Tylor's Researches into the Early History of Mankind. Tylor, whose older brother Alfred we encountered at Moulin-Quignon, was a close friend of his fellow Quaker, the banker and collector Henry Christy, whom he met on a tram in Havana, Cuba in 1856.[49]

Tylor was more interested in accounting for the myriad customs of the world's peoples and at the same time recognizing the common possession of language, laws, morals, religious belief, and the useful arts. These were probed for clues to history. The Stone Age made a late appearance in chapter eight in Researches, where he drew two conclusions: throughout the world

'Australians making flakes.

Figure 6.3 In later editions of *Pre-Historic Times* Lubbock introduced this image of stone knapping by Australians in the chapter on the uses of stone in ancient times. It was redrawn from an illustrated account by Thomas Baines (1866), 'the well-known African traveller'. Lubbock pulled back from captioning this a Palaeolithic scene, but the image is out there, open to that interpretation by the reader if they so wish. Hunters were now part of social evolution and a progressive world history.

Source: Lubbock 1869.

the use of stone preceded the appearance of metals, and within the Stone Age itself there was evidence of an upward direction. He disagreed with the recently deceased archbishop of Dublin, Richard Whately (1787–1863), who had made the case that 'savages' were incapable of improvement.[50] All the evidence assembled by Tylor pointed to the opposite. Human history could be *traced up* and 'the history of mankind has been on the whole a history of progress'.[51] Like Lubbock he did not believe that 'savages' were the modern representatives of deep history. But they did provide a space to think through the problem.[52] And when it came to explain change in customs and technology he identified three mechanisms; *invented* at home, *imported* either through the migration of peoples or diffusion of ideas, and *degraded* from a more civilized state.[53] The last mechanism he discounted in favour of the growth in time of Man's power over Nature.[54]

Here, then, were two ways to probe deep history, tracing back and tracing up the stream of time. Lubbock, as the anthropologist and historian George Stocking shows,[55] changed his mind between 1865 and 1870, when he published *The Origin of Civilisation and the Condition of Primitive Man* and adopted Tylor's history of tracing up. In the process he toned down his extreme negative views of 'savages'. The result was a much clearer evolutionist view of human history. Both John and Edward placed emphasis on mental evolution producing progressively more rational solutions to the problems of survival. Diffusion still played a part, as Tylor showed with his favourite example of the introduction of the Malay blast furnace into Madagascar,[56] but overwhelmingly Edward decided that 'the wide differences in the civilization and mental state of the various races of mankind are rather differences of development than of origin, rather of degree than of kind'.[57]

Lubbock's two books with their two approaches to deep human history ran together, never out of print up to his death in 1913 and each going through seven editions of constant updating. His readers always had a choice: trace history up or trace it back, depending on which book they picked off the shelf. What united the two tomes was his unwavering optimism about progress. Indeed, with each new edition of *Pre-Historic Times* John became more convinced in the progressive Victorian vision of the unity of the human race. In the third edition of 1872 he chirpily concluded that 'the habits and customs of savages while presenting many remarkable similarities which, as it seems to me, go far to prove the unity of the human race, still differ greatly, and thus give strong evidence of independent development'.[58] Even amongst 'savages' strivers will be found who make a difference, the motor of self-help driving developments in deep history. John's optimism, however, produced a sneer from Hunt's Anthropological Society in its review of *Pre-Historic Times*:

> We have one little problem to propose to Sir John Lubbock. Granted that the Somme has been at work for twenty thousand years in cutting its way down to its present bed; query, where will it have got down to by the time when these things shall have come to pass?[59]

Figure 6.4 The popular science writer Louis Figuier (1819–94) had Émile Bayard (1837–91) engrave scenes of Palaeolithic life in 1870. This scene depicts the ancient Somme gravels. Boucher de Perthes's axes are shown hafted as the heroic figures tackle a cave bear to protect a female cowering in the cave. Figuier accepted the time revolution but not the idea that species could change. In this he followed the leading scientific authority Armand de Quatrefages. Humans were a separate creation, and evolution from a lower form was, he thought, a degrading explanation of our origin. Bayard is best known today for his poster girl image of Cosette in *Les Misérables*, published in 1862.
Source: Figuier 1870.

For the Anthropologicals there could be no improvement in the savage races (Figure 6.4). Even unlimited time was insufficient to change their fixed racial character.

Stone Ages and Clubs 1865–6

But there was someone else who demanded Lubbock's attention in 1865: Charles Lyell. The charge of plagiarism brought by Falconer and supported by Prestwich had died away. Now it was revived by John. He had always thought that his fellow time revolutionaries had been treated poorly by Lyell, their work not given due credit. But at the same time he had asked them to tone down their attacks, and his review of *Antiquity of Man*, while critical, was civil.[60] This despite the matter of his 1862 paper on the Somme terraces, which John complained had been deliberately overlooked by Lyell.[61] Clearly something rankled and fuelled the slow burn of resentment. During the next two years and up to publication in 1865 strongly worded letters were exchanged in what Lubbock believed was a clear case of plagiarism by Lyell of his papers about the Danish shell mounds and the Swiss Lake villages.[62] Huxley and Hooker were brought in to mediate. The row seemed to have been patched up in private, so imagine Lyell's consternation when he opened the copy of *Pre-Historic Times* Lubbock had sent him to find the note shown in Figure 6.5 added to the preface.

Referencing the work of others was sparser in the 1860s than it is today, and historians who have undertaken a close analysis of the disputed passages think Lubbock was overreacting.[63] Lyell had rewritten passages without giving John due credit, but at the same time Lubbock had borrowed heavily, word for word, from the Swiss archaeologist Charles Adolphe Morlot (1820–67) with a similar lack of detailed acknowledgement.[64]

CHISELHURST,
February, 1865.

————————

NOTE.—In his celebrated work on the "Antiquity of Man," Sir Charles Lyell has made much use of my earlier articles in the "Natural History Review," frequently, indeed, extracting whole sentences verbatim, or nearly so.   But as he has in these cases omitted to mention the source from which his quotations were derived, my readers might naturally think that I had taken very unjustifiable liberties with the work of the eminent geologist.   A reference to the respective dates will, however, protect me from any such inference.   The statement made by Sir Charles Lyell, in a note to page 11 of his work, that my article on the Danish Shell-mounds was published *after his sheets were written*, is an inadvertence, regretted, I have reason to believe, as much by its author as it is by me.

Figure 6.5 Lubbock's preface in the first edition of *Pre-Historic Times*, 1865, accusing Lyell of plagiarism. Like a bad tweet, it was taken down in later editions. *Source*: Lubbock 1865.

The scientific circle of committed time revolutionaries and evolutionists was stiflingly close, as they visited, dined, and exchanged offprints of their papers. At times they found it difficult to remember where either information or a telling metaphor to describe deep history had come from. One of these comparing human migrations to the spread of weeds appeared in the first edition of *Pre-Historic Times*:

> there can be no doubt that man originally crept over the earth's surface, little, by little, year by year, just for instance as the weeds of Europe are now gradually but surely creeping over the surface of Australia.[65]

There was no acknowledgement, so the reader might think this was Lubbock's own expression. A very similar metaphor, however, had appeared in Alfred Wallace's paper read the year before in front of Hunt's baying anthropologists.[66] And at the same time Joseph Hooker was using the same colonial image in a paper in *The Natural History Review*.[67] When he recycled it six years later in *The Origin of Civilisation*, a more careful Lubbock now acknowledged where it came from.

So what was all the fuss about? The letters exchanged between Lubbock and Lyell grew nastier in their accusations and counter-accusations.[68] Was this unfinished business from the slight still felt by John's friends, Falconer and Prestwich? It mystified the circle around Darwin and caused much dinner table speculation. Hooker, who thought the whole affair far worse than Falconer's earlier outburst, claimed to have got to the bottom of it when he wrote breathlessly to Darwin in June 1865:

> And now my dear D. shall I tell you what is at the bottom of it all?— perhaps you won't believe it—it is just this—that Lady Lyell will not call on Mrs Busk nor invite the Busks to her parties. This the Lubbocks' & Huxley's resent.[69]

He went on to remind the recluse of Down House that the Lyells and the Busks both lived in Harley Street, London. When writing his *Antiquity of Man*, Charles Lyell repeatedly called on his neighbour George Busk to pump him for his knowledge of anatomy and fossils. Busk, a great friend of Huxley through their shared experience of working as ship's surgeons, had translated Schaffhausen's paper on the Neanderthal skull used by both Lyell and Huxley. In September 1864 he visited Gibraltar with Hugh Falconer to examine its limestone caves and inspect an ancient skull from Forbes'

Quarry. Found in 1848 by quarrymen, this adult female skull had been ignored when first reported by Lieutenant Flint to Gibraltar's Scientific Society. Perceptively, Busk saw its significance as another example of a Neanderthal.[70] But once his scientific uses were over, the Busks were cold-shouldered by the Lyells. The snub to his wife, Ellen Busk (1816–90), 'a most thoroughly accomplished clever person, excellent wife & mother, really scientific, & the kindest and most hospitable charitable person alive',[71] was keenly felt by Hooker and Lubbock, but more so by Huxley, who regarded her as his closest confidante. Besides her 'penetrating grey eyes', which he so admired,[72] she had the reputation of a freethinker and was not a member of the silent sisterhood; according to Huxley her words 'are sharp witted, spirited and playful. They can be sarcastic enough.'[73] Here was a woman unintimidated by alpha male scientists, but less enamoured to their wives. There is no evidence, however, that Lubbock's attack on Lyell stemmed also from an affection for the 49-year-old Ellen Busk that rivalled Huxley's and led him to attack the Lyell family's perceived ill-treatment of her.[74]

It remains difficult to explain John's 'evil spirit'[75] towards Lyell, just as Falconer's outburst seems an overreaction, hardly excused by leaping to the scientific defence of his milder friend Joseph Prestwich. There was just something about the 67-year-old Lyell which the 31-year-old Lubbock simply did not like. More to the point, Lyell had broken the unwritten geologists' code,[76] where honesty was vital to the pursuit of their stratified 'soul food'. Lubbock, however, soon saw the weakness in his case, and later reprints of *Pre-Historic Times* had the offending note removed. And it did not stop him sitting under Lyell's chairmanship and alongside his good friend John Evans on the Kent's Cavern Exploration Committee. Busk, who had brought the Neanderthal skull from Gibraltar to prominence, was not invited.

The tables around which the pot of resentment was stirred transmuted in 1864 into a new scientific dining club. In itself this event was non-newsworthy but the eight founder members who dined at St George's Hotel, Albermale Street, on 3 November combined a formidable array of scientific talent, linked by longstanding friendships and membership of the stately, and influential Athenaeum Club in Pall Mall.[77] Described by Hooker as himself, 'Huxley, Lubbock, and half a dozen others'[78] those invited to the inaugural dinner were Busk and Spencer, the muscular scientists Frankland and Tyndall, and the mathematician Thomas Archer Hirst (1830–92). At 57 Busk was ten years older than Hooker, while John was the youngest at 30. All except the Rt. Hon. John Lubbock were middle-class and several of them self-made professionals like Tyndall and Huxley. They all placed science above theology. The purpose of the club was to guide the growth and development of the body scientific.[79] They acted as a pressure group to shape the direction of learned societies and government policies.

A ninth member, another mathematician and friend of Tyndall, William Spottiswoode (1825–83), was invited to the next dinner. They never got around, as originally intended, to inviting a tenth friend and like-minded scientist as a permanent member. But that was to be expected of a club that in Spencer's recollection had no rules, except to have none.[80] They were also helped out over what to call themselves by Ellen Busk, who, no doubt tiring of the clever names endlessly bandied about, came up with the simple moniker of the 'X-Club'. This was her sarcasm at play, since 'X' is the Roman numeral for 10 and they had been unable to decide on who the tenth might be.[81]

The nine X-Club members formed three groups. Spottiswoode, Lubbock, and Busk provided not just intellectual *gravitas* but old money and the influence attached to it. The new scientific professionals, Huxley, Hooker, and Frankland, had limited means as well as young families, while the bachelors Hirst and Tyndall were free of these commitments but in the case of Herbert Spencer, often hard up.[82]

The X-Club dined on a Thursday before meetings of the Royal Society. By the time of the last dinner in March 1893, it had met 240 times.[83] Besides eating and talking they had, in the early years, summer outings to the country with their wives, the YVs of the X-men, as the joke went,[84] enlivened by Huxley reading Tennyson aloud.[85] Lubbock held open house at Lamas, and this continued when he moved back to High Elms, next to Down House, after his father's death.[86] Guests were invited to attend the dinners, and though never one of these, John Evans was considered as a permanent member after Spottiswoode died. But in true X-Club tradition they never got around to appointing him.[87]

The X-Club was most effective as a scientific ginger group. It revitalized and reformed learned societies—notably the Royal Society and the Linnean—and many of its members were elected to positions of president and vice president as a result of their letter-writing campaigns.[88] It also had an impact on the time revolution through one of its most significant successes, coordinating opposition to Hunt's Anthropological Society of London. This had broken away from the Quaker-based Ethnological Society with its roots in the older antislavery Aborigines Protection Society, an anathema of 'religious-mania' to the new Anthropologicals.[89] Race had split the Ethnologicals in 1863, when Lubbock became its president and Spottiswoode one of its secretaries. Many followed Hunt into the

Anthropologicals, and their membership grew quickly. They too had an exclusive dining club, 'The Cannibal Club', where members were called to order with a gavel shaped like a 'Negro's head'.[90]

As Hunt's racist, anti-evolutionary science gained traction the X-Club could not sit by and watch.[91] The issue of how to drive forward an evolutionary anthropology that embraced the time revolution and natural selection was too important. This was the new content for the new timescale of unlimited time. Moreover, many in the Ethnologicals, such as Huxley and Lubbock, supported the opposition to slavery.[92] They were helped by the fractious nature of the Anthropologicals, who were riven by bitter internal battles, unbridgeable rifts, debts, and financial impropriety.[93] Even so, some, like John Lubbock's intimidating future father-in-law Augustus Lane-Fox (1827–1900),[94] better known as General Pitt-Rivers after his promotion to major general in 1877 and name change on the death of his cousin in 1880, belonged to both societies until it was pointed out he must choose. Lane-Fox was a firm believer in historical progress, a mixture of Spencer and Darwin, a process that for him proceeded by bifurcation and continuity and where the artefacts made by hunters and gatherers in his large collection of ethnographic objects epitomized arrested growth, not degeneration.[95]

With these firmly held views Lane-Fox chose the Ethnologicals, as did Edward Tylor and the time revolutionaries John Evans and Joseph Prestwich.[96] Together with the X-Club they strived for a reconciliation which was made all the easier by the sudden death of Hunt in the summer of 1869. Without their repellent leader the Anthropologicals were a spent force. Huxley and Lane-Fox, as president and secretary of the Ethnologicals, pushed hard for a merger of the two societies and a change of name. To Lubbock's disgust, particularly because they asked him to be its first president, it was to be known as the Anthropological Institute of Great Britain and Ireland.[97]

Eight years of vituperative damage came to an end in 1871. Once inside the anthropological citadel the X-Club consolidated their position. Busk followed Lubbock as president in 1873 and two years later Lane-Fox took over, declaring in his presidential address of 1876 that 'the stern law of the survival of the fittest' had made reconciliation between his Anthropological Institute and the rival Anthropological Society unnecessary, as the latter was now extinct.[98] John Evans became president in 1877, and Tylor's election, which followed two years later, saw the decade out. This was how the X-Club and its fellow travellers kept an iron grip on the time revolution and the deep history of race.

Measures of Slippery Time

Deaths marked the end of the time revolution's decade. Uncle Hugh had overdone it on his return journey through Spain from Gibraltar. The route took George Busk and him to Cordoba, and then it was a journey of twenty hours in a mule drawn carriage over mountainous terrain en route to Madrid. Grace received a typically bravura account of overheated axles at the top of the Sierra Morena and an enforced stay in Don Quixote country, 'without anything but starved donkeys, lanky Dons, and pinched, scowling, inhospitable women'. They were getting hungry: 'Busk in a frantic moment of starvation impulse called for cheese, but something foul deluged with rancid lamp-oil was put before him on a dirty plate.' Uncle Hugh had had enough and resolved to walk the remaining forty miles to Santa Cruz de Mudela, where they could catch the train to Madrid. Instead, he caught a chill waiting for the repair to the axle.[99] They were marooned there for two nights, and, to compound their misery, on arrival in Madrid Busk had his pocketbook with all its geological notes stolen.

Falconer's illness got steadily worse after his return to London. He kept working, attending a council meeting of the Royal Society on 19 January to support Darwin receiving its prestigious Copley Medal, an honour the X-Club angled for. He came back depressed and feverish, and a rheumatic fever set in.[100] He still fired off notes to friends; Grace took dictation when he could no longer hold a pen. Joseph received one of his last notes: 'I would have seen you today if I could, but they would not let you come up.'[101] He died on 31 January 1865 and is buried in Kensal Green Cemetery.[102]

Henry Christy, two years younger than Hugh Falconer, died also in 1865 at the age of 55. He was visiting some new caves in Belgium with his friends the Lartets when he caught a severe cold that developed into inflammation of the lungs, and he died in France on 4 May. The originality of his preliminary contributions to their joint project into the caves of the Dordogne is striking, and it is one of the time revolution's greatest losses that he never finished the full work as planned. The nation, however, benefited from the bequest of his magnificent collection of antiquities and ethnographic objects. These went to the British Museum along with a sizeable endowment that his friend Augustus Franks put to excellent use in expanding and enhancing the Christy collection, which is one of the British Museum's most important.[103] Falconer commented that 'the instinct of a collector is to amass, hoard and retain', but in Christy's case it was also to share knowledge. He and Lartet distributed their finds from the Vézère around many museums, displaying in Falconer's view a 'higher impulse' that superseded the 'mere collector's instinct'.[104]

The Lubbocks, Nelly and John, narrowly escaped death on 9 July 1865, when the afternoon train from Paddington to Birmingham derailed after one of the engine's wheels came off at 60 mph. With the stiff upper lip of the muscular scientist he told his mother what happened,: 'the bumping got worse and worse, we were thrown backwards and forwards in the carriage, and though it seemed rather a long while, the only distinct idea I remember was that in a few minutes more we should probably solve many of those questions which interest us so much'.[105] Their first-class carriage was on its side in a field. John was unscathed, but Nelly was badly shaken and bleeding profusely. Helped by Annie, his wife's maid, unharmed in third class at the back of the train, he tied up Nelly's hand and arm: 'She was very brave, and declared that she was not hurt.' John could have said much more. Nelly was seven months pregnant with her sixth and last child, Rolfe, born on 19 September 1865. After this rail accident her health was fragile until her death fourteen years later.

The next year, and unexpectedly, Civil Prestwich died at the age of 44. She had devotedly supported Joseph's work and run their household at Kent Terrace for ten years. The day she died, 27 December, he wrote an anguished note to John Evans: 'I have lost the best of sisters. She passed away this morning tranquilly and without pain. I feel the loss is to me irreparable. She was my object in life, and so good, gentle, and affectionate.'[106] For some years he had been overworking and had suffered a stress-related illness.[107] At Civil's urging they had planned to move out of London to lead a quieter life. Above the village of Shoreham in Kent they found a large plot for sale, and for eighteen months before her death they had planned the house and gardens at Darent-Hulme (Figure 6.6).[108] The interior had intricate geological decorations, marble inlays, and depictions of his career's highlights: foliage of the coal measures and extinct animals, with flints in the outer walls.

The last time revolutionary to die that decade was Jacques Boucher de Perthes, at the age of 80 in Abbeville on 2 August 1868. He had enjoyed almost a decade of antediluvian vindication, first with the stone implements and human antiquity and then, at least in the eyes of many French scientists, with Moulin-Quignon. Now the scientific establishment wanted his collections, and at the invitation of Napoleon III he was asked to donate his prehistoric finds to the Musée d'Archéologie Nationale at Saint-Germain-en-Laye. A highly

(a)

(b)

Figure 6.6 A view of the Prestwiches' house, Darent-Hulme, on the chalk hills above the River Darenth, Shoreham, Kent (a), and the Dining Room with its ceiling, now whitewashed over, inspired by geological subjects (b).
Source: G. A. Prestwich 1899.

gratified Jacques said yes, as long as he could register them himself.[109] It took a little longer than anticipated, and pieces were still being catalogued almost fifty years later. What he bequeathed was problematic. There were many good pieces, but there were also forgeries by the *terrassiers* and a large number of the natural, figured animal stones that he alone was so convinced by. These were weeded out by the new curator of the museum and one-time political firebrand, Gabriel de Mortillet (1821–98).[110]

Greater glory was to come. In the year before he died, Boucher de Perthes's unpolished axes from Abbeville were given pride of place in the Galerie de l'histoire du travail (Gallery of the History of Work) in the Exposition Universelle held in Paris.[111] This huge world fair was an international showcase of France, and Jacques was at the heart of it.[112] The guidebook to the prehistoric rooms, which also included ethnographic artefacts, was written by de Mortillet. With a flourish embellished by his hatred of clerics Gabriel declared that the exposition demonstrated beyond any doubt that humans had a greater antiquity than any biblical chronology. He took the opportunity to bang the patriotic drum for French deep history, which demonstrated, in a single country, the great law of human progress from the most ancient to modern times.[113] Under de Mortillet's impetus the study of prehistory would become the science of human progress, complemented by another law, that similar historical developments in deep time are found across geographical space. This was a restatement of a universal history, the three sequential ages of Stone, Bronze, and Iron of the Northern Antiquaries rolled out on a world stage. Additionally, de Mortillet asserted, an analogy could be drawn between the oldest human cultures uncovered by Boucher de Perthes and those of modern 'savages' studied by ethnologists and displayed in the exposition for all the world to see.[114] This was what the Palaeolithic looked like.

The rest of Jacques's collection was left to Abbeville so long as it was housed in his home the Hôtel du Chépy and not rearranged for a hundred years.[115] He also left 10,000 francs to erect a simple but durable monument to himself in the Cimetière Notre Dame de la Chapelle overlooking the Somme Terraces.[116] The tomb is definitely durable but certainly not plain. On a plinth of white marble which lists other members of his family subsequently buried there,[117] together with his achievements, family crest, and honours—among them the Légion d'Honneur but not a fellowship in the Académie des Sciences—lies a sprawling bronze bed containing Jacques taking his last breath (Figure 6.7). A quill pen slips from his fingers, and by

Figure 6.7 The writer at rest, at last. Jacques's tomb in the Cimetière Notre Dame de la Chapelle, Abbeville. His brother Étienne and some of his nephews are also buried here.
Source: Author's photo.

his side are two of his books, his autobiography *Sous dix rois* and his greatest claim to fame *Antiquités celtiques et antédiluviennes*.

The optimism in France generated by the Exposition Universelle was short-lived. Three years later the French army surrendered to the Prussians after the battle of Sedan on the Belgian frontier. A disastrous campaign started in July 1870 and ended in ignominy less than a year later, to be followed by the insurrection of the Paris Commune, which ended in a bloodbath. The Second Empire collapsed, and Napoleon III, captured at Sedan, went into exile in Chislehurst, Kent, where he died and was buried in January 1873. During his short sojourn he visited John Lubbock's home at High Elms, where he saw the newly commissioned pictures of lively scenes of ice age life.[118]

Following Civil's death, Joseph's sister Emily now ran his household. Grace continued to live in London at her Uncle Charles's house, with visits to Forres to see other members of the extended family. Until he died, Falconer with Evans's connivance had tried unsuccessfully to marry Joseph off to some 'not-impossible Josephine'. Now this lady turned out to be Grace, a near neighbour on Regent's Park.[119] Their engagement was announced at the end of 1869, and they were married the next year on 26 February at St Marylebone Church, London. In a letter to Prestwich confirming him as the next president of the Geological Society, more evidence of the X-Club at work, Thomas Huxley compared matrimony to a swimmer's initial plunge: 'only don't be so mean as to go and tell a certain lady I said so, because I want to stand well in her books'.[120] They moved to Shoreham, and two years later Joseph retired from the wine trade, much to Grace's relief. Overwork and stress had surfaced again.[121] His appointment to the chair in Geology at Oxford in 1874 marked the full transition from bachelor and scientist workaholic to married professor and scientist at leisure; now supported by Grace, his 'most enthusiastic scientific friend, adviser, and co-worker'.[122]

Another transition occurred in 1870 with John Lubbock's election as the Liberal MP for Maidstone (Figure 6.8). It was his third attempt to enter Parliament. Support had come from the X-Club, and John Stuart Mill endorsed his candidature.[123] In the general election four years later he successfully defended the seat with a majority of sixty-six votes.[124] In the first four years of his parliamentary career he introduced four private member's bills that went on to become Acts.[125] The first, the Bank Holidays Act of 1871, was a trademark X-Club project, since its purpose was to give working people the leisure time to profit from education. This practical self-help freed up time to visit museums, libraries, and historical sites. Many condemned it as a licence for increased drunkenness. But seaside towns such as Margate applauded the move and bank holidays were known as St Lubbock Days.

The Cycle of Time

Darwin declared that the mist had been lifted from his eyes. Now he could understand what a million years really meant.[126] The person who had enlightened him was James Croll (Figure 6.9), a figure as socially and economically distant to Darwin as the geographical distance between Down House and Croll's modest home in Scotland. With characteristic simplicity this philosopher-scientist asked those, like Darwin, who were challenged by the enormity of a geological timescale to stretch a narrow strip of paper 83 feet 4 inches long around the walls of a large room.[127] That was the length of a million years. One hundred years, which Darwin's imagination could grasp, would then be a minute 1/10th of an inch snipped from the paper timescale.

Figure 6.8 John Lubbock in the House of Commons. The caricature by Sir Leslie Ward appeared in *Vanity Fair* on 23 February 1878.

Source: © National Portrait Gallery, London.

Figure 6.9 James Croll (1821–90) developed an astronomical theory of what caused ice ages to occur and recur. One friend recalled his lighter side:

> I saw Dr. Croll two days before he died. At that time he was very weak and exhausted, but mentally he was as clear and eager as in his best days. He was exceedingly anxious to discuss with me some of Mr. Herbert Spencer's fallacies, but his wife had warned me to stay only a very brief time with him. His effort to speak brought on a fit of coughing, to relieve which he was getting whisky in teaspoon measures. In that connection he made almost the only little joke I ever heard him utter. 'I'll take a wee drop o' that,' he said, 'I don't think there's much fear o' me learning to drink now! (Irons 1896: 488)'

Source: Irons 1896.

Of all those associated with the 1859 time revolution, James Croll was by far the most self-effacing. His low-key autobiographical sketch[128] reveals an unflinching commitment to a moderate Calvinism, as might be expected for someone born near Perth in 1821, and a need for solitude. He was the time revolution's Adam Bede, a stonemason's son who began his working life as a travelling millwright, but not he admitted a very good one, and then as a carpenter until an injury to his elbow made it too painful to work. He then tried selling tea in Elgin, which failed. Then, in 1848, and now married, the Crolls opened a boarding house in Blairgowrie which also failed. He turned to selling insurance, but ill-health continued to plague him, and he was not a success, mostly because 'study always came first, business second: and the result was that in this age of competition I was left behind in the race'.[129] Samuel Smiles, who resolutely believed that 'man perfects himself by work more than by reading',[130] would have despaired of James's inability to better himself in the Age of Equipoise.

But material success was not James's path. He believed in self-education, not self-betterment as measured by wealth, and was a serious metaphysical thinker as shown by his first book, *The Philosophy of Theism*, published in 1857. He argued from first principles rather than building a case from facts.[131] With such a mindset he was drawn to the 'hard' physical sciences to answer large questions in a 'softer' earth science such as geology. He never took an interest in human antiquity. The problem he concentrated his formidable reasoning upon was an explanation for the ice age.

The turning point in a life of struggles and setbacks came in the autumn of 1859, when Croll was appointed as a caretaker at Anderson's College and Museum, Glasgow.[132] The duties were light, the salary meagre but sufficient, and the library full of the scientific books that a knowledge-hungry scientist like James needed.

His starting point was the privately printed books by the geologist Louis Agassiz (1807–73) and the mathematician Joseph Adhémar (1797–1862), who had argued in the early 1840s for a glacial epoch (Agassiz) and looked to astronomy to explain it (Adhémar).[133] In geological terms their ice age was Lyell's mixed bag of difficult deposits labelled Pleistocene,[134] the Drift that Prestwich and Evans investigated to good effect for traces of human antiquity. And those deposits with their extinct animals were thought by them, and many others, to indicate a single cold period.

Croll's great insight was to suggest that a change from cold to warm conditions was precipitated by changes in the orbital path of the Earth around the Sun and in the wobble of the Earth around its own axis. The former is known as *eccentricity* as, over time, the Earth's elliptical orbit stretches and contracts. The latter is the *precession* of the equinoxes. When eccentricity was high, the Earth would be farther from the Sun and therefore colder. And when, due to precession, this coincided with the winter solstice taking place when the Earth was farthest from the Sun, the conditions were right, he argued, for glaciation to occur, not simultaneously across the globe but alternately in the northern and hemispheres (Figure 6.10). When there were ice sheets in the north, it was mild in the south, and vice versa. In

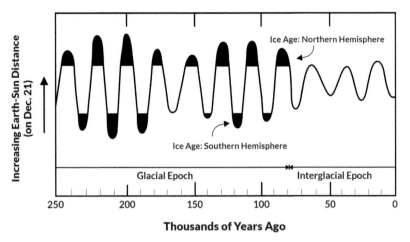

Figure 6.10 Croll's concept of successive ice ages in each hemisphere. These resulted from the knock-on effects of changes produced by alterations in the Earth's orbit, which determines the distance from the Sun. According to Croll, the period of ice ages came to an end 80,000 years ago, when warm, interglacial, conditions prevailed.
Source: Christian Hoggard, after Imbrie and Imbrie 1979.

James's words, 'the recurrence of colder and warmer periods evidently points to some great, fixed, and continuously operating cosmical law'.[135]

Astronomers including Sir John Herschel (1792–1871) and Urbain Leverrier (1811–77) had measured these changes and tabulated the results that James could now consult in the Anderson Library. Croll's love of the laws of numbers attracted him to these calculations, and after further computations he saw them as a possible cause for the ice age. But he was clear also that the Earth's orbit alone did not lead to ice sheets and glaciation. It was the indirect chain of events that such changes produced, principally in the displacement of ocean currents and the reflection of heat back into space due to extensive snow cover. In four papers published during the 1860s he set out his theory and then reissued it in definitive form in 1875 in his most important work, *Climate and Time*. What he overturned was the idea of a single unbroken ice age. In fact Croll's greatest contribution was to send geologists into the field to discover interglacials, those warm periods sandwiched between the cold ice advances that Lubbock had hinted at in his paper on the Somme terraces, only for it to be dismissed by Prestwich.[136]

The most important endorsement for Croll's multiple glaciations and interglacials came from the Scottish geologist James Geikie (1839–1915), who took on board his arguments in the groundbreaking *Great Ice Age*, which appeared in 1874. This book set the modern synthesis of Pleistocene glacial changes on its way. Using geological evidence Geikie argued for repeated periods of warm and cold, overturning the idea of a single ice age. Nor did Geikie's support for Croll end there. With his older brother Sir Archibald (1835–1924), they found Croll a job in the Geological Survey of Scotland in 1867 and, when well enough to undertake fieldwork, James played a small role in untangling the complexities of the Drift to ground-truth his theory of multiple glaciations.

For the time revolutionaries it was Croll's chronology that was important. Unlimited time was now in their grasp. To determine the timescale James started with the physicist Joseph Fourier's (1768–1830) calculations, based on the principle of cooling, that the Earth's crust was 98 million years old.[137] Then he took the tables of the Earth's orbital eccentricity and from these calculated that Lyell's Tertiary and Quaternary Periods[138] lasted 3 million years, during which there were five episodes of glaciation at a continental scale.[139] The two most recent ice ages coincided with the Earth's orbit in a phase of high eccentricity, producing alternating glacial climates in the northern and southern hemispheres between 980,000–720,000 and 240,000–80,000 years ago. Prestwich and Evans's glacial period, with its mammoths and cave bears, therefore, began and ended much earlier than they thought.

Joseph was always sceptical of these ages, maintaining until the end of his life that humans and extinct animals must be brought closer to the present. Such foreshortening was an attempt to reconcile his belief in the separate spheres of geology and human antiquity. Croll, the devout Congregationalist, avoided this question by never venturing a comment on human antiquity, except once, when, in passing, he let slip that the Mosaic chronology of 6,000 years since the creation of man was insufficient to account for the erosion of the landscapes once covered in ice.[140]

The follower of the time revolutionaries' bandwagon, Charles Lyell, was impressed enough to discuss Croll's theory in the last three editions of his *The Principles of Geology*.[141] Darwin included his ice age theory in the fifth edition of *The Origin* in 1869,[142] while Lubbock also used it in later editions of *Pre-Historic Times*.[143] Both Lyell and Lubbock latched onto the periods of high orbital eccentricity, which in the last one million years showed four pronounced peaks at 100,000, 200–210,000, 750–850,000, and 950,000 years ago. But deciding which peaks covered the antiquity of humans as dis-covered in Pleistocene/ Drift, deposits was a matter of judgement. Lyell originally favoured the 800,000 age for the extreme cold of the ice age, but between the tenth and eleventh editions of *Principles of Geology* he changed his mind.[144] Lubbock remained convinced through successive editions of *Pre-Historic Times* that the later date of 200,000 years ago fitted the evidence for erosion in the Somme.[145]

At the close of the time revolution's decade no one was closer to a definitive answer. But in the ten years since 1859 human antiquity had increased from 6,000 years to 100,000, and now that tentative figure pro-posed by Lyell before Prince Albert in Aberdeen had doubled. And in the background lurked the possibility, thanks to James Croll, that Lubbock's age of 200,000 years for human antiquity needed to be quadrupled. Ever cau-tious, John Evans did not speculate about numbers, but even so he thought it probable that humans were older than the Pleistocene, 'though', and here he showed considerable perspicacity, 'it will probably not be in Europe that the evidence on this point will be forthcoming'.[146]

Croll's ages were disputed in his lifetime and, in particular, when the last ice advance ended. North American geologists showed it took place a mere 10,000 years ago.[147] James's prediction that cold and warm periods alter-nated between the northern and southern hemispheres was also short-lived, replaced by multiple cycles of global ice ages and interglacials. Where he was right was in his insistence that changes in orbit and the precession of the equinoxes did not *directly* lead to repeated ice ages. Sitting at his caretaker's desk in the Anderson, he had correctly identified that explaining the com-plicated geology which contains the evidence for human antiquity depends on understanding the principles that drive our planet's orbital history. And with that insight an unlimited timescale was possible.

The treatment of Croll by the time revolutionaries was, despite their differences of opinion, class, and social standing, entirely to their credit. This might not have been the case. Prestwich and Evans liked inductive reasoning and, above all, evidence. Croll's deductions, which put principles above facts,[148] were contrary to their way of operating, and they distrusted an investigation that began with a theory. Even so, this scientific loner, humble museum caretaker, and devout Calvinist was never dismissed out of hand as a small-town savant. Quite the opposite. He was elected to the exclusive Royal Society, where entrance was restricted to fifteen new fellows a year, in 1876. His nomination was signed by Darwin, Lubbock, Tyndall, and Evans, and a group of Scottish geologists. When they were presidents of the Geological Society, Prestwich and Evans found funds in 1872 and 1876 to support Croll,[149] who could no longer work due to debilitating pains in his head.

Ascending the Stream of Time

Thanks to James Croll, time was moving inexorably backwards. The Scottish philosopher John McLennan (1827–81) put it bluntly in 1869: 'the antiquity of man is very great—the popular chronology', by which he meant Genesis, 'entirely wrong'.[150] De Mortillet agreed and organized the French evidence into Palaeolithic stages in the National Museum.[151] Progress in deep human history was now on permanent display.

By the close of the decade human antiquity was a done deal. In his introduction to *The Descent of Man* in 1871, which is more about the causes of racial variation and less about human ancestry, Charles Darwin declared, 'The high antiquity of man has recently been demonstrated by the labours of a host of eminent men, beginning with M. Boucher de Perthes; and this is the indispensable basis for understanding his origin. I shall therefore take this conclusion for granted'.[152] The time revolution was over. Darwin pointed his readers to the books by Lyell and Lubbock if they wanted more information.

Prestwich, Lubbock, and Evans had done their job. Joseph was soon to become professor of Geology at Oxford and set up house there with Grace. His days as a time revolutionary were largely done. John Lubbock was now first a politician and social reformer and second an optimistic evolutionary scientist. John Evans saw his post-revolutionary role as the steadfast complier of the archaeological facts that underpinned their proof of human antiquity. But even he, the great collector, wryly admitted that the subject of stone implements 'is one which does not readily lend itself to lively description, and an accumulation of facts...is of necessity dull'.[153] This was in 1872, when he published the bookend to the time revolution's long decade, *The Ancient Stone Implements, Weapons and Ornaments of Great Britain*. No expense was spared, with hundreds of woodcuts to illustrate Palaeolithic, Neolithic, and Bronze Age stone objects.[154] Throughout, John stuck to description. Speculations about the likely ages of these stone tools he left as usual to Lyell and Lubbock. But his huge testament to the importance and wonders of stone implements did fleetingly capture the challenge that he and Joseph had experienced together with John Lubbock,

Hugh Falconer, Jacques Boucher de Perthes, and all those other time revolutionaries as they followed the stream of time to the springs of human history:

> The investigators into the early history of mankind are like explorers in search of the source of one of those mighty rivers which traverse whole continents: we have departed from the homes of modern civilization in ascending the stream, and arrived at a spot where traces of human existence are but few, and animal life has assumed strange and unknown forms; but further progress is for the moment denied, and though we may plainly perceive that we are nearer the source of which we are in search, yet we know not at what distance it may still be from us; nor, indeed, can we be certain in what direction it lies, nor even whether it will ultimately be discovered.[155]

And that was the beginning of their legacy.

Notes

1. Falconer, Letter to *The Times*, 25 March 1864.
2. Lartet and Christy 1864: 235; Cook 2012.
3. Evans had seen something similar in 1860 from Reigate, England (see Chapter 4).
4. Lartet and Christy 1864: 253; Oakley et al. 1971: 136–7.
5. Lartet 1865–75: 206 and Plate XXVIII, p. 168. The find was also shown to Quatrefages and Desnoyers, who were part of the Moulin-Quignon Commission, and to Augustus Franks.
6. Christy 1865: 371.
7. Reprinted in the monumental book *Reliquiae Aquitanicae* funded from his estate. John Evans provided much advice, as did his friend at the British Museum, Augustus Franks, who had worked closely with Christy, a well-known collector. After Christy's death Franks was able to use the Christy fund to make many significant purchases for the British Museum, and he remains one of their greatest benefactors (Cook 1997, 2012).
8. Christy 1865: 372.
9. Stocking 1987: 50. See Chapter 5.
10. Trautmann 1992.
11. Christy 1865 [1875: 12].
12. A burial had been found during excavations by Dean Buckland in 1822–3 (Sommer 2004). The significance of this Upper Palaeolithic burial of an adult

male dismissed as a Roman prostitute and given the unfortunate moniker of the Red Lady of Paviland because of the ochre colouring over the skeleton, was overlooked (Oakley et al. 1971).

13. The Danish Kjökkenmöddings would now be classified as Mesolithic, while many of the waterlogged Swiss Lake villages are Neolithic in date.

14. Desmond 1998: 324; Wilson 2003: 253–7.

15. Hunt trained as a doctor and followed in his father's footsteps as a speech therapist.

16. Hunt 1863a: 1.

17. Stocking 1971: 379.

18. Hunt 1863b.

19. Hunt erroneously claimed longer arms, elongated heels, weak thumbs, and large molars.

20. Hunt 1863b: xvi.

21. Desmond 1998: 326.

22. Wallace 1864: clxv.

23. Wallace 1864: clxv.

24. Wallace 1864: clxviii.

25. Wallace 1864: clxxxvii.

26. The boys might have preferred stirring adventures of male camaraderie from the productive pen of R. M. Ballantine: *Coral Island* (1858), *The Gorilla Hunters* (1861), or *Away in the Wilderness* (1863).

27. Anything about the Irish, for example, was an open goal for his audience: 'When people live on poor vegetables instead of roast beef and plum-pudding, their jaws grow large, and their lips grow coarse, like the poor Paddies who eat potatoes' (Kingsley 1863: Kindle p. 1570).

28. Conlin 2011: 178.

29. https://en.wikisource.org/wiki/Hunt,_James_(1833–1869)_(DNB00). accessed 29 September 2020 Hunt also successfully treated Lewis Carroll in 1860 (Green 1953: 154).

30. Kingsley, Letter to Prestwich, 29 August 1859 (G. A. Prestwich 1899: 136).

31. Kingsley 1863: Kindle p. 1868.

32. Kingsley 1863: Kindle p. 1531.

33. Kingsley 1863: Kindle p. 1593. Du Chaillu also got a verse in *Monkeyana* as part of the argument between Owen and Huxley:

> The apes have no nose,
>
> And thumbs for great toes,
>
> And a pelvis both narrow and slight;
>
> They can't stand upright,
>
> Unless to show fight,
>
> With 'DU CHAILLIU,' that chivalrous knight!

34. Conlin 2014: Kindle p. 2748.
35. Evans 1872: 425.
36. Stocking 1987: 150–6.
37. Paddayya 2016. Bruce-Foote arrived in India in 1858 and made his finds of unpolished stone implements on 30 May 1863.
38. Lubbock 1872 (3rd edn): Figs 199 and 200.
39. Lubbock 1865: 490. He never wavered in this opinion, using it to close every edition of *Pre-Historic Times* from 1865 to 1913 (Murray 2009).
40. Lubbock 1865: 491.
41. Lubbock 1865: 491.
42. Lubbock 1865: 255.
43. Described by Stocking (1987: 153) as 'a grim Cook's tour of savagery'.
44. Tylor 1865: 195.
45. Murray 1992 provides a full discussion of the importance of this object and the normalization of the notion that Tasmanians were living representatives of the Palaeolithic.
46. These were standard views for 1865 and have to be compared with the argument that the savage races had degenerated, as propounded by the Duke of Argyll (1869), which Lubbock (1870) attacked.
47. Lubbock 1865: 445–7.
48. Lubbock 1865: 336.
49. Stocking 1987: 157; Larsen 2014: 13–36 covers Tylor's Quaker background and his loss of faith as his interest in anthropology grew.
50. Whately (1856) set out a theory of civilization that Tylor (1865: 160–2) quoted from at length in order to disagree with it.
51. Tylor 1865: 363.
52. Tylor 1865: 368–9.
53. Tylor 1865: 186.
54. Tylor 1865: 190.
55. Stocking 1987: 150–6.
56. Tylor 1865: 167–9. The two regions were connected by language, which indicated a migration from Southeast Asia to Madagascar.
57. Tylor 1865: 361.
58. Lubbock 1872: 557.
59. Anonymous review of *Pre-historic Times* in *The Anthropological Review* 1865: 346.
60. *Darwin Correspondence*, Lubbock to Darwin, 7 April 1863; see also https://www.darwinproject.ac.uk/lyell-lubbock-dispute, accessed 2 August 2020.
61. *Darwin Correspondence*, Darwin to Lyell, 17 March 1863.
62. Wilson 2002.
63. Wilson 2002; Bynum 1984.
64. Bynum 1984.
65. Lubbock 1865: 476.

66. Wallace 1864: 165: 'just as the weeds of Europe overrun North America and Australia, extinguishing native productions by the inherent vigour of their organization, and by their greater capacity for existence and multiplication'.

67. Hooker 1864: 124. 'In Australia and new Zealand... the noisy train of English emigration is not more surely doing its work than the stealthy tide of English weeds, which are creeping over the surface of the waste, cultivated, and virgin soil, in annually increasing numbers of genera, species, and individuals' (Lubbock 1912: 401).

68. Wilson 2002; https://www.darwinproject.ac.uk/lyell-lubbock-dispute, accessed 2 August 2020.

69. *Darwin Correspondence*, Hooker to Darwin, 2 June 1865. Patton presents the case that Darwin had put Lubbock up to it and then had to deny all knowledge to Hooker (Patton 2016: Kindle p. 73).

70. Busk 1865. If Busk had rescued the Forbes' Quarry skull from scientific obscurity sooner, the fossil ancestor we know as Neanderthal (*Homo neanderthalensis*) would have called *Homo calpensis*, after the Roman name for Gibraltar, and regressive behaviour everywhere would have had to be called Calpensic.

71. *Darwin Correspondence*, Hooker to Darwin, 2 June 1865.

72. Desmond 1997: 160, 341–2.

73. Wilson 2002: 78.

74. Wilson 2002: 79.

75. *Darwin Correspondence*, Hooker to Darwin, 2 June 1865.

76. See Chapter 1.

77. In 1864, when it first met, seven of the nine members of the X-Club were members of the Athenaeum. Hirst and Spencer were elected in 1866 and 1868 respectively (Reviewer 2 comments).

78. Barton 1998: 442.

79. Barton 1998: 442 and 2018 for a full account of the X-Club.

80. MacLeod 1970: 309.

81. MacLeod 1970: 309–10.

82. Barton 1998: 442 (Stocking 1971: 381).

83. MacLeod 1970: 318. Barton (2018: Kindle p. 9205) identifies three phases in the life of the X-Club. Its period of greatest influence, when its members pulled together, lasted fifteen years from the mid-1860s to the early 1880s.

84. Desmond and Moore 1991: 532.

85. MacLeod 1970: 313.

86. Barton 1998: 440–1.

87. MacLeod 1970: 313–14. It may be that Frankland blocked his election. He had a heated disagreement with Evans in the journal *Nature* in 1876 over water pollution by the paper industry.

88. For details of the far-reaching influence and importance of the X-Club, see Barton 1999; MacLeod 1970; Desmond 1998; Owen 2008, 2013, 2014.

89. Stocking 1971: 379.

90. Stocking 1971: 380.

91. Barton 1998: 437.

92. The X-Club and Huxley in particular waged a campaign against Governor Eyre of Jamaica, who had suppressed a revolt by black farmers in 1866 with great brutality (Stocking 1971: 380).

93. The leadership of the Anthropologicals was accused in 1866 in *The Athenaeum* of 'charlatanism, puffery and jibbery' (Stocking 1971: 382; Flandreau 2016). The two best-known Anthropologicals were the explorer Richard Burton (1821–90), who served as its president, and the poet Charles Swinburne (1837–1909).

94. Nelly Lubbock died in 1879, and on 18 May 1884 he married General Pitt-Rivers's second daughter Alice Fox-Pitt (1862–1947), twenty-eight years his junior.

95. Lane-Fox 1867. He illustrated his view with three papers on the evolution of warfare and its weapons (1867, 1868, 1869). In a series of developmental drawings he linked the Somme axes to much later finds in a classic progression from simple to complex. He did the same for the evolution of Australian fighting clubs and spears and drew inspiration from Evans's 1850 chart of changing coins (Chapter 3) and the idea that variation arises from copying.

96. Prestwich joined in 1869 just as the Anthropologicals' star fell from the sky (Reviewer 2).

97. Lubbock and Busk hated the term anthropology, preferring ethnology. Nelly Lubbock wrote to Emma Darwin in November 1873 to see if she could get her husband to add his name to the subscription to pay off a £700 debt left by the Anthropologicals which, as long as it stood, prevented a return to the name for the society which her husband and Busk preferred (*Darwin Correspondence*, E. Lubbock to E. Darwin, 29 November 1873). Despite Nelly's pleading nothing changed. The Anthropological Institute is now the Royal Anthropological Institute.

98. Gamble 2014: 149; Lane-Fox 1876: 487.

99. Falconer, Letter to McCall, 19 October 1864 (Milne 1901: 54–7).

100. Falconer 1868: 48–9.

101. G. A. Prestwich 1899: 195.

102. https://www.findagrave.com/memorial/21795/hugh-falconer, accessed 2 August 2020.

103. Cook 1997; Caygill and Cherry 1997.

104. Falconer, Letter to *The Times*, 25 March 1864, p. 10.

105. John Lubbock, Letter to his mother, Lady Lubbock, 10 July 1865 (Hutchinson 1914: 77–8).

106. G. A. Prestwich 1899: 203.

107. G. A. Prestwich 1899: 198.

108. The house was listed by Historic England in 1975, by which time the ceilings had been painted over; see https://historicengland.org.uk/listing/the-list/list-entry/1243748, accessed 2 August 2020.
109. Cohen and Hublin 1989: 227.
110. Richard 1999.
111. This was the second world fair held in Paris and was housed in a magnificent glass and iron oval building that covered the Champ de Mars (Cohen and Hublin 1989: 248–9).
112. De Mortillet 1867: 3. A flint axe from Menchecourt with a note attached by Boucher de Perthes was one of the exhibits; see https://en.wikipedia.org/wiki/Exposition_Universelle_(1867)#/media/File:Biface_de_Boucher_de_Perthes_MHNT.jpg, accessed 2 August 2020.
113. De Mortillet 1867: 184. See Defrance-Jublot (2011) for de Mortillet's views on the church.
114. De Mortillet 1867: 186–7.
115. Cohen and Hublin 1989: 232.
116. Cohen and Hublin 1989: 237.
117. His brother Étienne (1791–1871), who was director of Customs in Corsica and St Servan and St Brieuc in Brittany, and some of his children are buried in the tomb. He was the only one of Jacques's siblings to outlive him. See https://gw.geneanet.org/wailly?lang=en&p=jacques&n=boucher+de+crevecoeur+de+perthes, accessed 2 August 2020.
118. Owen 2013: 110. See Chapter 7.
119. Evans 1943: 142. Grace gives a detailed account of their honeymoon in southern Europe (G. A. Prestwich 1899: 216–25).
120. Huxley, Letter to Prestwich, 16 December 1869 (Huxley and Huxley 1900: 334).
121. Grace Prestwich, Letter to Evans, 8 February 1872, Ashmolean Museum, *Evans Collection*.
122. Woodward 1893: 246.
123. MacLeod 1975.
124. Owen 2013: 96. He stood twice, unsuccessfully, for the constituency of West Kent in 1865 and 1868.
125. Thompson 2009: Appendix 3.
126. *Darwin Correspondence*, Darwin to Croll, 19 September 1868.
127. Croll 1868c Part I: 375. Darwin used this example in the fifth edition of *The Origin*.
128. Croll in Irons (1896: 9–41).
129. Irons 1896: 31.
130. Smiles 1859: 3.
131. Finnegan 2012.
132. Irons 1896: 91.
133. Agassiz 1840; Adhémar 1842. Both books were published privately. For an excellent introduction to the history of ice age science, see Imbrie and Imbrie 1979.

134. Hamlin 1982.

135. Croll 1864: 129. See Fleming 2006: 47 and Imbrie and Imbrie 1979 for further explanations of the astronomical theory.

136. See Hamlin 1982: 573 for more details. See Chapter 5.

137. Croll 1864: 137.

138. Lyell's Tertiary covered the Eocene, Miocene, Older Pliocene, and Newer Pliocene epochs. The Pleistocene and Holcene are epochs in the Quaternary.

139. Croll 1868b Part II: 145, 1868c: 385.

140. Croll 1868a Part I: 382. See Finnegan 2012: n. 24.

141. Fleming 2006: 49. The tenth edition appeared in 1866 and the eleventh in 1872, where chapter 13 (pp. 272–97) was devoted to Croll's work. The final, thirteenth edition appeared after Lyell's death in 1875.

142. Finnegan 2012: 74.

143. The third edition of Pre-Historic Times (1872) has an extended discussion of Croll's theory.

144. Compare Lyell 1866: 295 and Lyell 1872: 287.

145. Lubbock 1872: 412–14.

146. Evans 1872: 426. In 1871, Darwin had argued, correctly, that Africa would be the birthplace for humans (1871: 199). Wallace (1864) in his address to Hunt's Anthropologicals had also suggested a tropical climate for human evolution from a primate ancestor.

147. Imbrie and Imbrie 1979: 95.

148. Finnegan 2012: 79. Croll was enthusiastic about Darwin and Wallace's natural selection because it was, in his metaphysical scheme, a directed force that was inexplicable without an active divine mind to propel it forward. Without this, natural selection was simply a mechanism that weeded out and was incapable of creating anything. Possibly, Joseph and Grace Prestwich would have agreed.

149. Irons 1896: 266 and 306. Prestwich used funds from the Geological Society's Wollaston fund and Evans its Murchison fund.

150. McLennan 1869: 272. He accepted an age for the earth of 100,000,000 years calculated by Sir William Thompson. Therefore geologists claiming 20,000 or 100,000 were not being 'greedy' (1869: 277). In this important essay he also discusses the primitive state of mankind and how to study human progress.

151. De Mortillet 1869: the Lower (Drift), Middle (cave), and Upper (abri) archaeology, with periods now named after artefacts dug up from Palaeolithic sites such as Saint-Acheul (Acheulean) and La Madeleine (Magdalenian).

152. Darwin 1871: 3.

153. Evans 1872: v.

154. Despite his title, the rest of the world did get a brief look in, as did Europe. But this was overwhelmingly a British volume of cave and river Drift implements.

155. Evans 1872: 425–6.

7

A Legacy of Zeal and Perseverance

As the century closed, Alfred Russel Wallace, one of its greatest scientists, pronounced on the successes and failures.[1] Singled out as great advances in *The Wonderful Century* were photography, telegraphy, electric light—safety matches and gaslight also received an honourable mention—and the Royal Mail, which, since 1840, had been delivering the post with a uniform charge irrespective of distance. On the negative side he denounced vaccination, 'a delusion' in his opinion, and the scientific neglect of phrenology, by which the shape of skulls can be read as the function of the brain within. Today Wallace has supporters among anti-vaxxers, while phrenology lives on amongst eugenicists and anyone who believes somebody's looks are a guide to their personal qualities.[2] Joseph's fine forehead would have got phrenologists buzzing with excitement, while John Evans's prominent chin would have confirmed for Wallace his steadfast nature.[3]

In a long list of scientific achievements the co-author of natural selection unsurprisingly devoted chapter XIII of *Wonderful Century* to evolution. Darwin had died in 1882 and was now being scientifically canonized. John Lubbock and other members of the X-Club had seen to it that the free-thinker of Down House was buried in Westminster Abbey. Wallace and four X-Clubbers were among the ten pallbearers.[4]

Geology received a chapter all to itself, where Wallace concentrated solely on the 'Glacial Epoch and the Antiquity of Man'.[5] Boucher de Perthes, Falconer, Prestwich, and Evans were all given due credit for an advance that in Wallace's opinion ranked among the most prominent examples of scientific progress in the nineteenth century.[6] He sided with the long chronology of Lyell and Lubbock and was relaxed about taking humans back into preglacial times, citing evidence from Burma and the gold-bearing gravels of California.[7] But as he was writing in 1897, the world's authority on stone implements, John Evans, was attending, as its president, a meeting of the British Association for the Advancement of Science held during August in Toronto.[8] North America was gripped by Palaeolithic fervour, with a bitter argument over the authenticity of ancient human remains and stone tools from a number of localities across the continent.[9] These were claimed to resemble the unpolished axes and stone flakes from Saint-Acheul and Hoxne and to match them in age. This was a balloon that Evans delighted in bursting, 'clearly and emphatically'.[10] None of the implements he was shown matched in any way those found on the Somme in 1859, or subsequently across the Old World, which everyone could see for themselves in the second edition of his *Ancient Stone Implements*, some advance copies of which he just happened to have with him.[11] After his pronouncement this phase of the American Palaeolithic Wars was over.

Many Legacies but Three in Particular

The time revolutionaries' demonstration of the high antiquity of Man was feted in its day, but what was their legacy? In this history of the 1859 time revolution I have until now tried to avoid hindsight by employing a rolling timeline of day, month, year, and decade to order the events of discovery and the process of acceptance. Hindsight is a brutal judge. We can see who was on the right track at a time when conviction and informed speculation was all they could hope for. But once the time revolution became a done deal for geologists and archaeologists, it is appropriate to loosen the stays on the corset of chronology. As Lubbock showed in his books, it is possible to write history by tracing it down as well as up.[12]

Through their zeal and perseverance, the principal time revolutionaries supplied the initial contents for a deep human history—stone artefacts, extinct animals, an ice age, unlimited geological time, speculations about age, and a face for the past drawn from the ethnographic present.[13] This content fuelled their three legacies, personal, scientific, and historical.

Their personal legacy can be measured by what they left behind: portraits, memorials, houses, families, and their collections (Figure 7.1). None of them is an A-lister from the nineteenth century. They don't rank alongside Huxley or Darwin, Hooker or Lyell. You won't find their portraits hanging in the National Portrait Gallery, London.[14] But that is what makes them interesting and important. They push their way into the story through their zeal and perseverance—well, maybe Lubbock had a silver-spoon introduction to the life scientific, but Jacques, John, and Joseph certainly did not. Their personal legacies are statements about how others saw them as well as their role in creating a new understanding of humans and time.

Figure 7.1 Transformed by time, the old, successful time revolutionaries preserved in oils. (a) Sir Joseph Prestwich, aged 83, artist unknown, painted from a photograph in the last year of his life. It was presented to the Geological Society by his wife.

Source: © The Geological Society of London.

Figure 7.1 (b) Sir John Evans, aged 82, painted by John Collier. The painting is in the Ashmolean Museum, Oxford. Several of his later portraits omit his glasses.

Source: © Ashmolean Museum, University of Oxford.

Figure 7.1 (c) Lord Avebury, Sir John Lubbock, aged 77, painted by Hubert von Herkomer. The painting hung in Bromley Museum, now closed.

Source: Bromley Libraries and Heritage.

The scientific endeavours of the time revolutionaries turned on an oxy-moron; a new science of the very old: prehistory. This science was their second legacy, and Evans and Lubbock started filling it with hard evidence. Ever since, archaeologists have populated earliest prehistory with three great armies of stone: the Lower, Middle, and Upper Palaeolithic. Many flinty battalions can now be found mustered under the banner of Lubbock's Old Stone Age, differentiated by regional and local names. These distinctive traditions of knapping flint within a set of common themes are of core importance to archaeologists as they order Palaeolithic data by time and geography to chart small but important technological changes. Their task is now much easier because reliable ages exist for the stone tools and these slot into a detailed picture of the fluctuating ice ages when ancient humans lived. Thanks to the time revolutionaries, archaeologists now sit alongside geolo-gists as scientists of the deep past, contradicting the idea that the study of the past is solely about ancient civilizations; its subject matter, as Evans had hoped, now broadened to include the objects of unwritten history.

The third and most significant legacy of the time revolutionaries was the release of history from the shackles of recent time. This possibility needed a geological timescale, but it did not lead to geology with humans added. It is possible, however, to see a different emphasis between geological and his-torical archaeologists, as identified by the historian of science Bowdoin Van Riper. The former are interested in the cultural boundaries drawn up by those flint armies. The latter see their purpose as defining national identities within national boundaries. With this distinction in mind Van Riper con-cludes, 'historical archaeology traced the history of a single people, geo-logical archaeology, that of all Europeans'.[15] Both types of archaeology can, however, only be understood in the historical context that shaped the time revolution. Lubbock and Evans reflected their country's interest in empire, where deep history in the form of an imperial Palaeolithic, found like the British in every continent, provided continuity and legitimation for the political order they imposed and enjoyed. In that sense they were geological archaeologists.

By contrast, historical archaeology will always start, as Macaulay and Freeman did in the nineteenth century, at some critical juncture in the national story. Points of origin such as 1066 or 1688 are not, however, readily available on the timescales of the time revolution. Gabriel de Mortillet, writing in 1897, knew this.[16] Confronted by the humiliating defeat of the French army at Sedan in 1870 and the annexation of Alsace by Germany, what concerned him was the identity of France. Here the role of deep time, when French history started in the Palaeolithic, provided the warranty of national unity at a time of grave existential threat.[17] As the historian of science Nathalie Richard points out, only a prehistorian like de Mortillet could 'appreciate and make evident that the roots of the French past are incredibly deep, a depth by which the current weight of national sentiment is to be measured'.[18] And that brings de Mortillet closer to a Northern Antiquary like Worsaae and the uses he made of prehistory to underpin Denmark's emerging national identity.[19] A strip of clear blue water exists between their use of deep history for the nation and their English contemporaries, Lubbock and Evans, who symbolize Britain's international power.

The time revolution had a unifying potency for the nineteenth century.[20] Its deep history served imperial rule and national sentiment. Today, an interest in empire has been replaced by the implications of prehistory's global reach for understanding human diversity as the outcome of our evolutionary history. This was hinted at by Evans,[21] and his legacy lies at the root of explorations into at least three distinctive forms of history described, as we shall see, by the adjectives, universal, big, and deep.

Today we may lack John Lubbock's cheery belief in the steady, upward progress of humanity. He was a child of the Age of Equipoise and benefited socially and economically from his century. Others were less fortunate (see Figure 2.7), so that drawing the lesson of progress from history, however old, has always been unconvincing. The history made possible by the time revolutionaries continues, however, to align a philosophy of change within a context of hope, and this is their greatest legacy.

Personal Legacies, Memorials, and Family

The events of 1859 did not raise the profile of Boucher de Perthes, Evans, Prestwich, or Lubbock much beyond the fields of archaeology and geology, although France looked after Jacques's posthumous reputation rather better than England did for the other three. The imposing exterior of 10 Kent Terrace, London, home to the Prestwiches for many years, bears no official blue plaque to commemorate one of the nineteenth century's leading geologists, let alone an original time revolutionary. Instead a later resident, E. H. Shepard, the illustrator of Winnie the Pooh, has that honour. Oxford, where Joseph and Grace lived for some years, has done better.

The city erected a black plaque in Prestwich Place, a cul-de-sac in the western suburb of Botley—their house in Broad Street was demolished to make way for the highly ornamented Indian Institute in 1883—to commemorate a professor of Geology 'influential in gaining improvement to the poor drainage of New Botley in the 1880's'.[22] The practical geologist would have approved, but there is more he could be publicly celebrated for. His friend and mentor Hugh Falconer is the only British time revolutionary to have his own museum, in Forres, his birthplace.[23]

Evans is lionized by Palaeolithic archaeologists and numismatists.[24] A small display celebrating his achievements can be found in the Ashmolean Museum, Oxford, which received much of his huge and surprisingly uncatalogued collection.[25] But the size of the Ashmolean's gallery space devoted to his son, Sir Arthur Evans, suggests that his greatest achievement was as father to the boy who long ago buried his dolls in the garden of Nash House.[26]

Once he enters politics in 1870, John Lubbock becomes a fringe player in the great evolutionary story. He evolves into Darwin's diplomat in the survival of the nicest.[27] A useful antidote to Huxley's ankle-snapping bulldog, as he was when he spoke at the famous Oxford debate in 1860,[28] but a background rather than limelight figure in the history of science. His collections went to the British Museum and the local museum in Orpington, now swallowed by Bromley, that was closed in 2015 because of government cuts, a bitter irony for an educator like Lubbock.

By contrast Jacques, or at least his bust, can be found looking down from the walls of several museums of archaeology and natural history, celebrating *le père de la Préhistoire*. There is a Musée Boucher de Perthes in Abbeville,[29] but not as he requested in his will in the Hôtel de Chépy, where he once lived. Abbeville and Amiens might have been on the front line of Palaeolithic research but they were overrun in the twentieth century by two world wars. Abbeville was shelled during the Great War, but Jacques's home escaped, only for his prized collection to be destroyed twenty years later when Rommel's army advanced across northern France in May 1940.[30] A further casualty was his fine bronze statue (Figure 7.2) by Emmanuel Fontaine, erected in his honour by fellow Abbevillois in 1908. De Crèvecœur would have been heartbroken.

All three Englishmen were publicly honoured. Evans was knighted in 1892 and so became the first 'Knight of the Spade'. After a lobbying campaign by his friends, Joseph joined the geological 'Knights of the Hammer' six months before he died on 23 June 1896. Lubbock had inherited

ABBEVILLE - La Place du Pilori
et le Monument Boucher
de Crévecœur de Perthes

Figure 7.2 Boucher de Perthes's statue was unveiled in the Place du Pilori, Abbeville, on 7 June 1908, forty years after his death (Manouvrier 1908). He gazes down at an unpolished antediluvian stone axe from the gravel pits of the town. The statue, his museum, and many other buildings in the town were destroyed on 20 May 1940 as the German army first pounded it with artillery and then dropped incendiaries.

Source: Author's collection.

his knighthood, but that changed in 1900 with his elevation to the House of Lords, where he took the title of Lord Avebury. For the previous twenty years he had been overseeing legislation, aided by his father-in-law General Pitt-Rivers, to protect the ancient monuments of Britain.[31] The Neolithic stone circles at Avebury and Stonehenge were two of the first monuments to be officially protected for the benefit of the nation.

The list of honours and positions held by all three time revolutionaries is quite overwhelming. One will suffice. Evans retired as manager of Dickinson's in 1885 at the age of 62 and devoted himself to long stints as trustee of the British Museum, treasurer of the Royal Society, president of the Society of Antiquaries and the BAAS and many county positions, among them Justice of the Peace and deputy lieutenant for Hertfordshire, serving as its high sheriff. He was granted the liberty of St Albans and is remembered for safeguarding the water rights of the county.[32]

Joseph continued his roles among the learned societies that he had belonged to for so many years.[33] While in Oxford, he wrote two large textbooks,[34] and when he fully retired to Shoreham in 1887, his workload was carefully supervised by Grace, whose object was to reduce his agitation to avoid the recurrence of illnesses that plagued him all his life. As might be expected, Grace judiciously curated his legacy to present posterity with a geologist almost too good to have walked the shifting sands and gravels of the Pleistocene Drift. Her work culminated in commissioning a flattering portrait in oils, now in the Geological Society (Figure 7.1), and a biography of her husband drawn from his correspondence and her memories.[35] She cannily played the Scottish card—how Uncle Hugh would have loved it!—and asked Sir Archibald Geikie to summarize Joseph's career. He proclaimed that Prestwich was revered 'as one of the last of the old heroic race of geologists', honoured for his perseverance, enthusiasm, scrupulous caution, infinite patience, and the exhaustiveness of his researches.[36] And set alongside these Smilean virtues, Geikie continued, his scientific zeal was matched by the charm of his personality and his genuine goodness.[37]

Lubbock had by far the busiest public life, combining a very active political career with a huge output of educational and uplifting books, many of which, like *The Pleasures of Life*, became bestsellers. Unlike Joseph, his immediate legacy was poorly served by Horace Hutchinson (1859–1932), a successful golfer and prolific author of sporting books, who rushed into print the year after John's death with a two-volume work of his selected letters and achievements that never rises above the superficial for such a 'busy' man.[38] Lubbock had to wait until the twenty-first century to be taken seriously.[39] Although never forgotten, he was never fully appreciated in the intervening century, a case of being Darwin's shadow.

John Evans fared only slightly better. There was no comparable life and letters immediately following his death. He had to wait forty years for his daughter Joan's family memoir, *Time and Chance: The Story of Arthur Evans and his Forebears*, where he is used to set the scene for her half-brother. Page-wise the memoir splits 40 to 60 per cent in favour of the son.[40] It was not until the centenary of John's death that the first detailed study appeared of his multifaceted achievements, scientific, industrial, and public, and his role as an enlightened ambassador for international cooperation.[41]

Joseph and Grace had no children. John Lubbock had eleven from his two marriages, none of whom predeceased him. But two followed soon after: Harold and Eric Fox-Pitt from his second marriage to Alice, died during the Great War aged 30 and 24.[42] None of his children continued his passion for anthropology, human antiquity, and the causes of the ice age.

Of Harriet Evans's five children, two died before their father, Alice in 1882 and Philip in 1893. Two of John's children followed his antiquarian pursuits and collecting passion. Arthur became famous for excavating Knossos, while Joan Evans was a celebrated medieval historian and an avid collector, like her father.[43] Her upbringing had that 'sternly archaeological' aspect which Fanny had experienced on their honeymoon in 1859, as did Maria, his third wife and Joan's mother, in 1892, when he took her on an equally romantic honeymoon to the gravel pits of Abbeville and Amiens.[44] As an example of John's perseverance and zeal, Joan was taken as a 7-year-old to visit those same gravel pits of Saint-Acheul. She recalled fifty years later the *terrassiers* in their 'deep-belted baggy corduroy trousers' who sold stone implements to her father that she treasured ever after: 'I doubt if any of them—my father included—remembered on that sunny August morning that in that gravel pit a new science of prehistory had had its beginnings.'[45]

All three time revolutionaries marked their success in the world with large new houses, none of which acts today as a 'shrine' to their memory. Joseph and Grace left Oxford and moved back permanently to Darent-Hulme in Kent. The house is now subdivided, its carefully chosen geological motifs hidden.

The large house that Evans built for Maria in 1906 at Britwell on Berkhamstead Hill and named after his birthplace in Buckinghamshire is now part of Castle Village retirement home, where it is known as the Mansion. The move from Nash House was a mammoth undertaking. So heavy was his collection, four tons of flints alone,[46] that when he sent it to his son Arthur, it broke the bed of Dickinson's largest lorry. His coins, always his first love, went with him to Britwell.[47] The forbidding mansion of his bullying uncle, Abbots Hill, overlooking Dickinson's paper works at Nash Mills, has become a girls' school. All the buildings on the Dickinson's site have vanished, and Nash House, vacant for some years and a ruin, was demolished to ground floor level and then rebuilt for residential use.[48]

The peerage seems to have gone to Lubbock's head. As befitted his new status, he bought, in 1902, the neo-Gothic pile of Kingsgate Castle, perched precariously on the cliffs at Broadstairs, Kent.[49] Renovating this dilapidated eighteenth-century folly took up a good deal of time and a great deal of money. Today it is subdivided into thirty-one apartments, protected from the sea by massive steel pilings. The Lubbock family home at High Elms was destroyed by fire in 1967. A short walk away, Down House, with its famous Sandwalk, is a place of pilgrimage for Darwin tourists.

Their businesses have all gone. Joseph Prestwich & Son was sold when he moved to Darent-Hulme. The year after Lord Avebury died, Robarts, Lubbock, & Co was merged with Coutts & Co., and Lubbock was no longer a name on a banking house. In 1914 it was a going concern with reserves of £500,000 and balances of £4 million.[50] Coutts became part of the Royal Bank of Scotland Group in 2000. It was the failure of RBS in 2008 that led to a government buyout of banks that would have appalled John Lubbock, the City grandee and advocate of free trade.[51] As for Evans's business, Dickinson's was a brand which at one time produced Basildon Bond notepaper. In 1926 it had a network of thirteen overseas companies.[52] The name has disappeared in England thanks to repeated mergers.[53] But the brand still thrives in Trinidad and Tobago, one of the original international companies. Throughout the West Indies you can buy school exercise books with 'Winners, A John Dickinson Product' printed on their marbled covers.[54]

In his will Prestwich left his great friend Evans six dozen bottles of his best Château Rothschild.[55] Evans reciprocated with an obituary that pleased Grace. She died three years later on 31 August 1899, having finished her labour of love, Joseph's Life and Letters. Her description of his last days when he was very weak is poetic, reminiscent of the death of the abused boy Smike in Nicholas Nickleby, surrounded by friends under a spreading tree on a shaded summer's lawn.[56] Two years later, Grace's sister Louisa produced a memoir of her indomitable sibling, beloved niece of Uncle Hugh, companion, and time revolutionary.[57]

As to their memorials, none matched Jacques's theatrical bronze effigy on a bed of marble in the Abbeville cemetery (Figure 6.7). Grace and Joseph were buried beside each other in the Shoreham churchyard of St Peter and St Paul. Inside, a stained glass window of Joy, Creation, and Love, completed in 1903, commemorates Grace, Joseph, and his sister Civil, his two companions in geology and the time revolution.[58] John Evans is buried with Harriet and Fanny in a simple stone grave in the churchyard of St Lawrence's, Abbots Langley. Sir Arthur Evans would join them in due course. On the wall of the church is a lengthy inscription listing John's many achievements, and prominent among these is his role in the time revolution, declaring him 'Amongst the first to demonstrate the vast antiquity of man'.[59] He died on 31 May 1908, and the mourners packed the streets of the small village.

Lubbock wrote Evans's obituary and was the last of the time revolutionaries to die. He kept writing to the end, arguing with all his Liberal values against the European war which now seemed inevitable and for the importance of free trade. He died of heart failure when returning to High Elms from Kingsgate Castle on 28 May 1913. The staunch ally of Darwin and the X-Club is buried at his chosen place of worship, St Giles the Abbot in Farnborough village, Kent, next to High Elms.[60] The grave was marked by a tall Celtic stone cross carved with bronze axes and the inevitable megalithic tomb.

Scientific Legacy

Prestwich never wavered in his interpretation of the geological evidence for human antiquity. The year before he died, he gathered some of his last papers into a volume tackling 'some controverted questions of geology'.[61] His great complaint was against the strict adherents to Lyell's principle of uniformitarianism, where the present is the key to understanding the geological past. What Prestwich questioned was not the principle itself but the English application of it. He challenged Lyell's view that in geology there was uniformity in both the processes of change, for example mountain building or the erosive power of rivers and ice, and the rates of those changes, or, as Joseph put it, in kind and degree. Without uniformity in processes between the present and the past there could be no geology. But what Prestwich favoured was European scepticism concerning uniformity in the rates of those changes. Here the present was not an infallible guide. Why, he asked, should the small time window of the present through which geologists measure the degree of change apply to all previous geological periods? Prestwich warned that if present rates of change were applied, geologists were blindly applying a rule that made them march to 'the martinet measure of time and change'.[62] As an instance of such unthinking application, he derided estimates for human antiquity of one or two million years.[63]

Prestwich had not changed his views in forty years. He remained convinced there was only one ice age, unbroken by warmer periods, and in terms of age everything should be brought forward rather than pushed back. His final word on the Drift, which by 1895 was increasingly referred to as the Pleistocene, reiterated his evidence-based view that it could be divided into three parts with the following ages:

- the *early glacial*, with humans at sites such as Saint-Acheul in the high gravel terraces of the Somme, dated to 47,000 to 38,000 years ago;
- the *full glacial*, as marked by ice advance and maximum cold and without traces of human presence;
- the *post-glacial*, when the ice melted and humans returned, dated to 10,000 to 8,000 years ago. The low level gravels of the Somme which he described in 1859 were the final stages of these meltwater floods.

But try as he might to hold back the tide, his views, based as they were on so many field visits to quarries, cliffs, and caves, had been replaced even before he died. James Geikie, spurred on by Croll's theory of multiple climate changes, had by the third edition of *The Great Ice Age* in 1894 amassed an impressive worldwide body of Pleistocene data. In northern Europe, where the ice sheets had their home, he had geological evidence for at least four major ice ages interspersed with three interglacials and a fourth in which we currently lived.[64] He mapped the extent of the ice advances through the moraines of northern and Alpine Europe and logged stratigraphic sections that showed interglacial sediments and the warm-adapted animals within them. He also determined that those loess sediments which Prestwich attributed to annual floods were more often accumulations of windblown soil, stripped from the dust bowls in front of the ice sheets, and increasing in thickness on a eastern traverse across the continent.[65]

Joseph had lost the argument even among his fellow time revolutionaries. Evans wanted more time for human antiquity but studiously avoided giving an age,[66] while Lubbock, addressing the British Association in 1881, summarized Croll's model of changing Earth orbit and announced 300,000 as the date of the last glacial epoch.[67] Gabriel de Mortillet came out in support of the long chronology, settling on Croll's figure of 240,000 years old for the implements of the Somme.[68] Before Lubbock died in 1913, the number of European ice ages based on the Alpine evidence would converge on the figure of four, separated by three warm periods.[69] Palaeolithic evidence was bit by bit slotted into this framework, which was thought to cover 600,000 years.[70] What it did not incorporate was James Croll's notion that ice ages resulted from changes in Earth's orbit, Lubbock's endorsement was discounted.

For many years Europe's archaeologists and geologists went out and found evidence of four ice ages and three major interglacials. Their colleagues in North America searched for a similar number. These were difficult field data, as Prestwich well understood. The Drift was the slipperiest of all geological periods, and the erosive power of water and ice meant that the evidence was always piecemeal. The geologist, as Geikie showed so well in his synthesis, had to piece together small segments of the geological story from one area and somehow join them to the sequence from another, matching bits of river terraces with raised beaches and moraines to cave deposits and ancient lakes. Pleistocene geology will always be a science, like archaeology, that depends on fragmentary evidence. That can make it difficult to separate out the framework for interpretation from the evidence. In 1859 Joseph took pride in his data-led approach, and Evans followed the lead of his older friend. Lubbock liked data but was open to big ideas, such as Darwin's and Croll's, that challenged the status quo.

In the absence of any independent chronology it is unsurprising, therefore, that by the turn of the century the Pleistocene was lumped into four main glaciations. Here was a framework derived from close observation of field evidence which then became the guide for the collection of more evidence. The framework and the evidence confirmed each other. The archaeologist's job was to act like a geologist and slot the stone tools and human fossils into the 'correct' glacial or interglacial. And once these were slotted home, the business of describing change could proceed. The causes of change were rarely investigated. A nod towards either natural selection or Herbert Spencer's mantra that simple inevitably leads to complex was sufficient.

The synthesis of such fragmented data and the archaeology they contained was remarkable and continued for the first half of the twentieth century. It was blown away, however, by a new archive of climate data that came not from the bits and pieces of the land but from the floor of the ocean.

Wiggle Curves and Climate

The discovery that the ocean floor was covered in fine sediments, the benthic oozes, had been made by H.M.S. Challenger during its three-year voyage between 1872 and 1875.[71] Oozes were dredged from the deep and analysed for their content. But the full implications for Pleistocene geology had to wait until a piston corer was invented that could retrieve a long column of sediment. This technical issue was solved in 1947,[72] and the finely stratified cores were then studied for the marine creatures they contained, tiny forams, to give a changing list of species and their abundance. The oozes built up continuously on the ocean floor, providing oceanographers with a seamless rather than fragmented record of the past. Then came the clever bit. The tiny carbonate skeletons of the same foram species could also be analysed for the ratio between two oxygen isotopes they absorbed while alive. These isotopes have different 'weights', the lighter O^{16} and the heavier O^{18}. When plotted against the depth in the core, this ratio between O^{16} and O^{18} produced a wiggle curve with peaks and troughs (Figure 7.3). This was the start of the evidence which proved James Croll right.

By the 1970s the deep-sea record had replaced the land-based framework of four glacial periods. The wiggle curve presented a slowly accumulating, continuous account of the Pleistocene. The crucial insight, made in 1967 by Sir Nick Shackleton (1937–2006), was that the ratio of the two differently weighted isotopes measured the changing size of the oceans and not temperature, as previously believed.[73] When sea level is high during the warm interglacials, then they are enriched by the lighter isotope O^{16}. As sea levels fall due to cooling temperatures, then these 'lighter' isotopes are drawn off to build the continental ice sheets, leaving an isotopically heavy ocean, enriched with O^{18}. A core, such as the one in Figure 7.3, taken in one place is measuring the global balance of ocean to land. There is nothing piecemeal in the archive of these wiggle curves, as they show the stately progression of Pleistocene oscillations from large to small oceans, warm to cold phases, interglacials to glacials and back again.

A Second Time Revolution

Unlocking the potential of these seamless records of climate change needed a second time revolution. It came after the Second World War as a spin-off from the study of radioactive particles to make atomic bombs. In 1948 the Nobel Prize-winning chemist Willard Libby (1908–80) realized that the orderly decay of radioactive isotopes such as carbon-14 could be used to date organic materials. Radiocarbon dating measures isotopes which have been decreasing at a fixed rate since the death of the animal or plant that

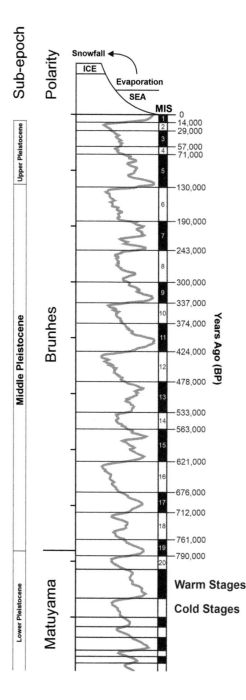

Figure 7.3 A deep-sea history of the last 800,000 years of changing climates, as recorded in Core V28-238 drilled in the Pacific Ocean and published in 1973 by Nick Shackleton and Neil Opdyke. The wiggle curve of O^{16} to O^{18} shows small, isotopically heavier global oceans to the left during cold phases and larger, isotopically lighter oceans to the right during warm phases. Odd and even numbers denote warm and cold phases respectively. These are known as Marine Isotope Stages, and cores like this one form the backbone of Pleistocene chronology, climate and environmental research.

The slow sedimentation rates in this core make it possible to read off age against depth, much as Leonard Horner tried to do in Egypt with the build-up of Nile silts. At a depth of twelve metres in the core is the magnetic reversal to what we understand as normal polarity, the Brunhes Epoch, dated on land to 787,000 years ago. This marks the internationally agreed boundary between the Lower and Middle Pleistocene. Eight pairs of warm and cold stages (e.g. Stages 19 and 18 and the most recent 5 and 4-2) can be seen in the Middle Pleistocene.

Source: Christian Hoggard after Gamble 1999: Figure 4.2.

contains them. As radioactive isotopes go, the half-life of carbon-14 is quite brief at 5,730 years. This means that radiocarbon can be used to date back to 40,000 years ago before the amounts of carbon become too small to measure. The spectacular Upper Palaeolithic finds made by Lartet and Christy, including an engraving of a mammoth (Figure 6.1), from the *abris* along the Vézère fell within this range. Radiocarbon could also date the organic sediments at the top of the deep-sea cores. But this left most of the cores undated, as well as archaeological sites such as Saint-Acheul and Hoxne from the Lower Palaeolithic.

Radioactive isotopes with much longer half-lives were needed, and among these the ratio of potassium to argon, K-Ar dating, is so great that it can be used to date the Earth's oldest rocks. This technique, however, is unsuitable for the sediments in the deep-sea cores, but on land it is ideal for volcanic rocks of all ages, allowing a smart triangulation to take place to date the wiggle curves. Lava dated by this method can also be measured for its magnetic properties. For reasons that are still obscure, there was one significant moment 787,000 years ago when the Earth's magnetic field reversed; the North Pole became the South Pole and vice versa. Known as the Brunhes-Matuyama boundary, after two pioneers of palaeomagnetic studies,[74] this same reversal in the Earth's magnetic poles shows up in the deep-sea cores. Once this triangulation was made from a well-dated magnetic signal on land to the same signal in the cores, it was possible to read off the ages of the warm and cold phases indicated by the wiggle curves. And this is possible because of the steady, continuous build-up of sediments on the ocean floor which contain the tiny marine forams that once lived at its surface.

The results overturned a century of speculation. Where there were once four ice ages and three interglacials in the entire Pleistocene, there are now eight pairs of glacial-interglacial stages in the last 800,000 years alone. Before this there were twenty-two weaker cycles, without major continental glaciations, back to 1.6 million years ago, when polarity changed again.

With an unbroken record available, attention turned back to Croll and the Serbian mathematician Milutin Milankovitch (1879–1958), who had made it his life's work to prove the astronomical theory of ice ages.[75] He calculated changes in solar radiation at different latitudes on the basis of the eccentricity, obliquity, and precession cycles known from astronomy. These three cycles, the stretch, wobble, and roll of the Earth as it spins around the Sun and its own axis, have periodicities of 100,000, 41,000, and 23,000 years. They interlock like gears in a gearbox, and the result, as laboriously calculated by Milankovitch, predicts a pattern of ice age and interglacial conditions that ebbed and flowed simultaneously, and not sequentially, as Croll believed, in the northern and southern hemispheres. The predicted Croll-Milankovitch climate cycles were now compared with the wiggle curves of the oceanographers. The match was obvious. Milankovitch was proved correct because the deep-sea curves using those oxygen isotope values changed according to the predictions that Croll had laboriously initiated a century before in *Climate and Time* and that Milankovitch had refined. The changes in the wiggle curves from a cold to warm phase could be attributed to the changes in the Earth's orbit and axial tilt.

More recently, the terraces which Prestwich and Lubbock started investigating have been tied into the ocean chronology.[76] The staircase of terraces and the varied sediments they contain have been described for both the Thames and the Somme, their ages confirmed by independent science-based dating. Lubbock's paper in 1862 on the Somme terraces, where he recognized fluctuations between an interglacial and glacial climate and correctly identified how fluvial terraces are formed, establish him as a pioneer in the mysteries of river histories.[77] Prestwich's fixed views on their formation, stemming from his firm belief in a single glacial period, have not weathered well.

And so, finally, to the age of the Palaeolithic archaeology in these deposits. The Saint-Acheul gravels from which the 1859 finds came are placed in Stage 12 of the deep-sea record,[78] dated to between 478,000 and 427,000 years ago (Figure 7.3). Hoxne falls in the following warm period, Stage 11, 427,000 to 364,000 years ago.[79] The fought-over finds at Moulin-Quignon, some of which are genuine, are among the oldest in the high terrace of the Somme, placed in interglacial Stage 15 with an age of 600,000 years.[80]

While oceanographers were transforming the Pleistocene, archaeologists using radiometric dating techniques were pushing back human antiquity out of the Pleistocene and into the older Pliocene period. The breakthrough came in 1961, when attention turned to dating a thin volcanic horizon, or tuff, that runs, like jam in a Victoria sponge cake, through the deepest levels at Olduvai Gorge, Tanzania,. This tuff yielded an age using potassium-argon techniques of just under 2 million years.[81] In Bed I of this great sequence, the archaeologists Louis (1903–72) and Mary Leakey (1913–96) had excavated stone tools and the remains of fossil humans. Evans would have been pleased. He confidently predicted in 1897 that the earliest humans would be found outside Europe, although he favoured India and East Asia as the cradle of humanity.[82] No doubt he was influenced by the find, in 1891, of a skull cap and femur from the island of Java, Indonesia. Its discoverer, the Belgian geologist Eugène Dubois (1858–1940), called the remains *Pithecanthropus erectus*, the 'upright-walking apeman'.[83] The Leakey's wrenched that focus away from Asia to Africa, where it has remained ever since.[84] Currently, the oldest stone artefacts, different from those from Saint-Acheul or Hoxne because there are no hand-axes, come from the site of Lomekwi in Ethiopia and date to 3.3 million years ago.[85] At this time there were no fossils close to *Homo* either in body shape or size of brain. Instead, the likely makers were the small-brained Australopithecines, 'southern-apes', whose upper body was still well adapted for climbing trees, but below the waist they could walk. Evans, the 'excavator' of *Archaeopteryx*, would have relished the evolutionary mixture of features.

Joseph Prestwich also argued for a preglacial age for humans, but he would never have countenanced 2 or 3 million years. The evidence was delivered to his front door at Darent-Hulme by Benjamin Harrison (1837–1921), the village shopkeeper in nearby Ightham village.[86] Harrison had been collecting stone implements from the high plateau gravels of the North Downs in Kent. He took the elderly Joseph in 1879 to see the sites, and the time revolutionary was convinced by these strangely shaped implements.[87] Among the genuine stone implements were some that were even more roughly made than those from France, Brixham Cave, or the rivers of England. These were dubbed eoliths, 'dawn stones', a term introduced by Gabriel de Mortillet in an attempt to push France's Palaeolithic farther back in time.[88] Joseph pointed out that nothing in Boucher de Perthes's collection compared to these artefacts 'of rude make and of peculiar type',[89] made by a 'simple and nascent intelligence'.[90] Here was another level in his terrace staircase from lower to higher gravels now topped by the plateau gravels. Height was always his measure of age.

That eagerness not to offend had led Joseph to befriend another scientific outsider, and there are similarities in his relationship with Harrison to that with Boucher de Perthes.[91] But championing the Kentish eoliths is reminiscent of his support for the exaggerated claims of Moulin-Quignon rather than Saint-Acheul and Menchecourt. Evans was dismissive of the evidence. There were genuine implements among Harrison's collection, but they came from all periods, Neolithic as well as the Palaeolithic. As for the roughly-made dawn stones with rolled edges, they were natural but not forgeries.[92]

In 1912 Steadfast Evans was not on hand to dispel Piltdown, the most infamous Palaeolithic forgery. Perhaps its perpetrator was familiar with Evans's role in the Moulin-Quignon Affair and waited until he was dead before trying his luck. The announcement by the solicitor and weekend archaeologist Charles Dawson (1864–1916), another scientific outsider, of a human-looking skull from Piltdown in East Sussex with an ape-like mandible received wide, and mostly uncritical, scientific support.[93] These remains were found together with an odd assortment of artefacts, stone and bone, one of which was fancifully dubbed the 'cricket bat', but nobody got the joke. Lubbock saw the material at the British Museum in February 1913[94] but raised no objections, unlike his immediate reaction to Moulin-Quignon. Instead he included it in the last and seventh edition of *Pre-Historic Times*, which appeared after his death.[95] The tools were of less significance than the human remains, which were given the name *Eoanthropus dawsoni*, 'Dawson's Dawn Man' to go with the 'dawn tools' he was found with. The remains established that human evolution proceeded first with the enlargement of the brain followed by the evolution of the teeth. Most importantly, Piltdown suggested that Britain played a key geographical role in the physical evolution of humans, and the brain evolved in Europe, not Africa.

As a result, Piltdown blocked the acceptance of the first Australopithecine found in a limestone quarry near Taung, South Africa, and named in 1925 by Raymond Dart (1893–1988).[96] The Taung skull completely contradicted its Sussex rival: the teeth in this fossil were leading the way, while its braincase had not started to expand much beyond what might be expected for a chimpanzee. Such was the power, however, of Britain's Piltdown that the significance of Africa's Australopithecines was ignored for some years.

A strong whiff of scientific imperialism kept Piltdown alive for much longer than it deserved. The fraud was finally discredited forty years later in 1953: the skull shown to be a few hundred years old and the jaw belonged to an orang-utan, its teeth filed down and stained.[97] Prestwich and Evans might have said it was the blinkers of theory obscuring the merits of the evidence. Unlike Moulin-Quignon, Piltdown was never subjected to a commission of inquiry, and if it had been, some reputations might have been saved. The stratigraphy was unconvincing, and the appearance of eolithic artefacts suspiciously coincided with the visit of someone who needed to be impressed. John would have seen through it at once. And I like to think Joseph also, no doubt after a bit of twitching brought on by not wanting to hurt Dawson's feelings. It is the sort of hoax that would have delighted SENEX[98] and even more the loathsome James Hunt of the Anthropologicals, as it illustrated for them the silliness of scientists who believed in human antiquity as well as those who traced connections between apes and humans.

Piltdown remains an embarrassment but not for the time revolutionaries. Their positive legacy carried on internationally. One campaign of the American Palaeolithic Wars may have been halted by Flint Evans in Toronto in 1897. But the search continued for genuine Pleistocene-age humans in North America. When the evidence came, it mimicked many of the experimental protocols seen in 1859, when John and Joseph went to Amiens. North America finally gained a deep human history at Folsom, an out-of-the-way-place in New Mexico in 1927. In a small arroyo, the excavator Carl Schwachheim (1878-1930)—and his scientific backers in Washington and New York, was rewarded with an aesthetically pleasing and skilfully made stone projectile point lying among the ribs of an extinct species of bison. Here was stratigraphic, faunal, and photographic proof, as in 1859, of Pleistocene antiquity.[99] When the site was eventually dated by radiocarbon, and then redated by the same technique at the end of the 1990s, it only just made it into the Pleistocene, with an age of 12,437 years.[100] When exactly humans arrived in the New World remains controversial, the subject of the latest campaigns in the Palaeolithic Wars,[101] but it is less than 25,000 years ago[102] and late in the story of when humans settled the earth.[103] Evans was correct in discounting anything that claimed to be as old as Hoxne or Saint-Acheul.

The legacy of 1859 is still strong in Palaeolithic archaeology. One of those lithic regiments beneath Lubbock's Palaeolithic banner is the Acheulean, named after the type site of Saint-Acheul and known by its distinctive hand-axes, or bifaces, one of which Prestwich dug out of the Fréville pit on 27 April. The Acheulean is one of humanity's most remarkable phenomena. Acheulean hand-axes are found from northern Europe to the coast of southern Africa and eastwards across Arabia and India and just into China. The rest of Asia, Australia. and the Americas are Acheulean-free, either because they were settled much later or by people who did not make bifaces, favouring instead a flake and stone core technology rather than cats' tongues and *coups de poings*, hand-axes. The Acheulean also defies gravity by lasting for over a million years. It was made by at least four species of the genus *Homo* over a time period when brains were increasing significantly in size.[104] Such stasis in technology confounds us with our expectations of incessant innovation inherited from the Victorians.

But some things do not change with further research. Almost a hundred years after Evans first published the flint bible, *Ancient Stone Implements of the British Isles*, a full gazetteer of all the finds made subsequently was compiled by a Cambridge research student, Derek Roe (1937–2014). He marshalled 100,000 artefacts from around the museums of Britain and devised a classificatory system for thirty-eight of the larger collections based on subtle measurements of hand-axe shape and subjected these to computer analysis. The result confirmed Evans's classification made after only a few days in France and written down on Blue Dickinson notepaper for the Royal Society: there were two groups of stone implements, collections dominated by either oval or pointed forms.[105] And like his predecessor, Roe did not conjecture on the reasons why this should be, but ever since his gazetteer has been an invaluable source for informed speculation.

The Legacy of Deep History

It is a convention, widely observed, that any history of Victorians needs a quote from Lewis Carroll's, *Alice*. The time revolution in human antiquity passed her by, but not time itself: 'For, with all her knowledge of history, Alice had no very clear notion how long ago anything had happened.'[106] Time shaped her Victorian readers. But like Dickens who opens *Our Mutual Friend* with 'In these times of ours, though concerning the exact year there is no need to be precise',[107] there is a cultivated vagueness about the calendar and the age of things.

The time revolutionaries are no different. They get the dates wrong as they remember events that happened in a rush. Charles Pinsard even wrote down the wrong year, 1857, for his scrapbook of the events of 27 April 1859 at Saint-Acheul. Jacques's great *Antiquités* bears the date 1847 but didn't appear until two years later. Joan Evans muddles up the dates of letters and moves her father's visit to Hoxne with Joseph forward by a year,[108] while John is confused about when he found a Roman coin at Fréville's pit. These inconsistencies, equivalent to errors in page proofs, do not make them bad chroniclers or unreliable witnesses of the history they created. Ironing out these discrepancies is an unimportant exercise, so long as the general movement from day to month, year, and decade runs its course. Their imprecision about the chronology of the present is understandable when viewed against the difficulties of estimating the age of the humans whose flint proxies they had recovered.

These small fallibilities make 1859 a very human history embedded within a deep history of humanity, a story of discovery and acceptance, circles and societies, friendship and clubs, sisters, wives, and uncles, the lifespans of the principals dwarfed by the vast lapse of ages inhabited by our remotest ancestors. They were outlived by the institutions that provided the intellectual scaffolding around the scientific edifice of human antiquity: the Royal Society, the Society of Antiquaries, and the Geological Society. Their houses may have been burnt down or changed use, their graves rarely visited and their collections closed to the public, but they are remembered as time revolutionaries through their papers in scientific and popular journals and through active research by archaeologists and historians who trace back to these intellectual ancestors.

To understand the significance of the events of 1859 I have presented them as the history of a time revolution. However, none of those who did the scientific heavy lifting in 1859 would have recognized the concept. Indeed, their journey from discovery to acceptance is more like a rapid courtship with the deep past than a revolution with time.

The time revolutionaries were clear about their achievement. It was to create a science of primeval man, of humanity's prehistory. This new science was an archaeological bridge between geology and anthropology, a means for others to link the great question of race and racial origins with a deepened sense of time and the stone implements of those times. It was a science because it depended less on a single act of discovery than the ability to repeat the experiment.[109] Hoxne repeated the experiment first undertaken in the gravel pits of the Somme. Then the numbers of Palaeolithic stone tools from France, Britain, Europe, and soon the world kept repeating the experimental findings that human antiquity had to be measured by geological timescales and that Boucher de Perthes's axes served as a valid proxy for the elusive bones of fossil ancestors.

Lubbock's science of prehistory fused nineteenth-century concerns about race with the need to rethink it once geological antiquity was accepted. Arnold Toynbee (1852–83) exhorted historians in 1884 to 'pursue facts for their own sake', which would have gone down well with Evans and Prestwich, 'but penetrated with a vivid sense of the problems of your own time', which would have appealed to Lubbock.[110] The time revolution played out against the immediate background of the Indian Rebellion (1857–8) and the American Civil War (1861–5). The challenge to colonial authority and the right to hold slaves ruffled the Age of Equipoise far more than the discovery of stone tools with extinct animals. But the demonstration of great antiquity would be part of the aftermath of both world-changing events. Certainly, the time revolution was a challenge to religion and Mosaic history but more to the Evangelical movement than the established church, as revealed by today's attitudes in the United States and Britain, where hostility in the former is balanced by the latter's neutrality to questions of human antiquity and teaching evolution.

After their deaths the study of primeval man became a quest for the evidence of human origins through time and across space. This was the enduring legacy of the time revolutionaries. When and where was the geologist's alternative to 'Eden'? What faces would the anthropologist find there? How did a simple stone technology measure the changes on the road to becoming properly human? Lubbock's Palaeolithic was the starting point for these origins. His Old Stone Age came to be defined by what it was not—civilization with cities, arts, architecture, writing, and constant technical advances—and never could be. Our Palaeolithic origin was a place to escape from, as enthusiastically supported by H. G. Wells (1866–1946) in his influential world history serialized in 1919.[111] Fuelled by unlimited time and the Smilean and Lubbockian virtues of ingenuity and application, the individual and the society they belonged to would, one day, arrive at the city on the hill. The lesson of prehistory continued as a lecture in progress.

Wells, rather than archaeologists, shaped the perception of our deep-time ancestors. In his short story The Grisly Folk published in 1921 he savaged Neanderthal, described with neutral language in 1863 by Huxley and Schaaffhausen.[112] Now this ancestor was to be feared with its 'big face like a mask, great brow ridges and no forehead, clutching an enormous flint, running like a baboon with his head forward and not, like a man, with his head up'.[113] When Neanderthals encounter true men, Homo sapiens, they are put to the spear in an evolutionary mercy killing. From his pen, deep human history became an excavation into the barbaric psyche that civilization had evolved to suppress.[114] Someone had to be blamed and who better than a fossil ancestor who couldn't answer back?

Wells was responding to Toynbee's sense of contemporary problems where the atrocities of the Great War had unleashed the beast in humanity; scratch the surface of civilization and, Jekyll and Hyde-like,[115] a primitive biological past takes control (Figure 7.4).

Ever since 1859 artists, and then film-makers, have depicted scenes from deep history and drawn inspiration from the raft of archaeological discoveries but equally from the iconography of Western art to visualize the primitive other.[116] Between 1869 and 1871 Lubbock commissioned his own view of Palaeolithic and Neolithic life from Ernest Griset (1844–1907), the artist of the ghoulish.[117] Nineteen pictures show scenes of deer and bison hunting and other stirring scenes of Palaeolithic life.[118] Realistic they are not, closer in sentiment and execution to Bayard's

Figure 7.4 An American recruiting poster from 1917 designed by Harry Ryle Hopps (1869–1937) captures the fears of writers like H. G. Wells, who transferred the 'animal' savagery of the battlefield to Neanderthal ancestors. Germany is shown as a gorilla, which, having destroyed Europe, is now on American soil. The bloody club which became the trademark symbol of the caveman, bears the slogan *Kultur*, a German concept being ridiculed in this image.

Source: Author's collection.

engravings in Figuier's *Primitive Man* (Figure 6.4). One is an ice age version of Landseer's *Monarch of the Glen*, painted in 1851, and now given an icy Pleistocene setting. Another shows a horde of tiny spear-wielding figures attacking a mammoth and being tossed about on its tusks in a comical rather than threatening scene of ice age life. No hand-axes are on view, and the paintings which Lubbock approved trivialize the Old Stone Age.

John never used these paintings to illustrate later editions of *Pre-Historic Times*. Instead, they hung at High Elms, where they were seen by the exiled Napoleon III.[119] One was intended as a gift for Darwin but never sent. After his death they could be seen in all their ordinariness in Bromley Museum.[120] But there were more. Prestwich had his own collection, now lost, of 'quaint geological pictures' by the same artist at Darent-Hulme.[121]

The time revolutionaries, and Lubbock in particular, would have been appalled by Wells's presentation of their evidence. But neither would they have recognized the rehabilitation of primeval man, again in the shape of Neanderthal, after the Second World War by William Golding (1911–93). In *The Inheritors* (1955) we are the nasty Pleistocene lords of the flies, and they are a source of innocent humanity.[122] The restoration culminates in Jean Auel's *Clan of the Cave Bear* (1980), the first in her Earth's Children series, with its feminist heroine Ayla. Here it is males rather than Neanderthals who are an evolutionary dead weight.[123]

Even though human origins exist as hard archaeological evidence, they nonetheless provide a bottomless pit of time into which all the undesirable and, occasionally, more noble elements of humanity can be dropped (Figure 7.5). This pit can be plumbed for explanations, or excuses, for ills against humanity perpetrated by individuals or nations, races, and genders. But not all is gloom in this imaginary abyss of the ancestral soul. The same pit can be examined for more optimistic messages about the present supported by the same hard evidence of stones and bones.

Universal and Big History

Prince Albert was one such optimist, when, in 1850, he expressed his belief that the destination of history was the unity of mankind.[124] A century later and his universal history seemed a step closer in the aftermath of world war, genocide, and holocaust. Archaeology's moment on the world stage came in 1948, when UNESCO drew up the *Universal Declaration of Human Rights*, boldly stating that:

Figure 7.5 Continued

Figure 7.5 Human origins have always been too important to leave to scientists alone (Hammond 1982; Moser 1998). Our remote ancestors have oscillated between the good and bad guys of human history, as shown by images of the same Neanderthal skeleton from La Chapelle-aux-Saints, France. Excavated in 1908 it was first seen as a brute, club-wielding Pleistocene mugger by František Kupka (1871–1957) in the *Illustrated London News* for 1909 (a).
Source: *L'Illustration* 1909.

Two years later, the same skeleton received a sympathetic treatment in Amédée Forestier's (1854–1930) picture, captioned 'Not in the Gorilla Stage' (b).
Source: *Illustrated London News* 1911.

All human beings are born free and equal in dignity and rights. They are *endowed* with reason and conscience and should act towards one another in a spirit of brotherhood (Article 1).

and the *First Statement on Race* two years later:

The unity of mankind while firmly based in the biological history of man rests not upon the demonstration of biological unity, but upon the ethical principle of humanity.

—which refers back to the earlier Article 1.[125] Such a universal document, while not always successful in practice, as subsequent wars and oppression have shown, could not have been written without the science of primeval man and a home for human origins in deep history. Our human endowment mentioned in Article 1 had to come from the deep history of humanity, where unlimited time had shaped reason and conscience.

The time revolution unlocked the possibility of an optimistic universal history. The three technological ages of the Northern Antiquaries, which Evans and Lubbock largely ignored, is one example. Their contribution was to add a deep basement beneath the Stone Age—Lubbock's Palaeolithic peopled with Evans's flint implements and aged by Prestwich's geology.

These three ages were further transformed in 1935 by the archaeologist Gordon Childe (1892-1957), fresh from visits to Stalin's Russia. He proposed, and found his proposal readily accepted, that the Stone, Bronze, and Iron Ages equated with two major revolutions in human history.[126] First came the Neolithic revolution, with agriculture, polished stone, and settled village life. Then, in the Bronze Age, the urban revolution ushered in the ancient civilizations of Egypt and Mesopotamia culminating in the classical worlds of Iron Age Greece and Rome. Inspiration for these dramatic historical changes came from the eighteenth century's Industrial Revolution, popularized by Toynbee in 1884, presaging the modern world. The two older revolutions confirmed the Palaeolithic as either inside human history and therefore its origin point or lying outside it, where its inhabitants belonged to the realm of natural history. Either way it was peopled by 'savage' hunters that Childe, an Australian by birth, had no time for. His motor for the revolutionary staircase was the upward curve of population pressing on resources and calling forth beneficial innovation in economy, technology, and society.

This historical engine was different from the Darwin-Wallace mechanism of natural selection. That might apply to the Palaeolithic and its brutal struggle against the inimical forces of ice age climates, but not to Neolithic famers liberated from a hunting existence and Bronze Age urbanites living on the labours of others. Their endowment, acquired in deep history, allowed them to escape the dictates of biology and the forces of nature. Lubbock and Evans would have approved. Childe's universal history, confined exclusively to the evidence from European prehistory and written before the radiocarbon revolution, has many nods to Spencerian progress, Lubbockian optimism, and Evansian pragmatism. Childe's message in 1940, as the shells were raining down on Boucher de Perthes's Hôtel in Abbeville, was that Europe had seen this before in the Bronze Age. Then, it had resisted 'Oriental despotism' by behaving, as Childe put it, in a 'distinctively European way'.[127] The European Urban revolution emerged stronger and better than that in the East because Europe fostered liberties founded on the freedom of its Bronze Age craftsmen. John Evans would have allowed himself a large Smilean grin.

Universal history made sense of the time revolution. Eventually the Palaeolithic shed some of its savage image and caught up. It did this by proposing its own human revolution positioned well before the economic upheavals of the Neolithic. Timing, as usual, is all. Some archaeologists place the revolution in the symbolic explosion of the Upper Palaeolithic, 40,000 years ago, when evidence for cave art, decorated objects, and ornate burials is commonplace.[128] Others point to the Eurocentric nature of this 'revolution'. Africa and, increasingly, Asia have much older hard evidence for the widespread use of symbols. Accordingly, the human revolution that wasn't a revolution at all starts at least 200,000 years ago outside Europe and, like one of Joseph's wines, took time to mature.[129] And now there is evidence that

Neanderthals were not so dumb but regularly made and used decorative symbols long before they had their Jean Auel—rather than H. G. Wells—encounter when *Homo sapiens* entered Europe.[130] But even that is not old enough for some. Another age for the Palaeolithic's human revolution takes it back to a million years ago and the appearance of larger brains, upright walking and dextrous hands for making stone tools and kindling fire.[131]

Lubbock would have engaged with these arguments, as no doubt would Boucher de Perthes. Meanwhile, John and Joseph would have shrugged them off as too much theory and returned to accumulating facts. They might have preferred instead the genre of big history that starts with the single cell and, sticking to Spencer's mantra, ends with the complexities of industrial society, a prime example being the science popularizer Henry Knipe's (1855–1918) long illustrated poem *Nebula to Man* published in 1908. Having rhymed his way through all the geological periods, Knipe finally arrived at ice age Europe:

> Primeval men are now upon the scene,—
> Short, thick-boned hairy beings of savage mien,
> With ape-like skulls; but yet endowed with pride
> And power of mind to lower brutes denied.[132]

The sweep of big history that charts the human place in the universe has been a favourite ever since with popular science writers.[133]

Big history underpins the current interest in renaming the most recent slice of geological time as the Anthropocene, when humans turned the tables and shaped geological processes. The accelerating pace of glacier retreat, desertification, rain forest reduction, and sea level rise bear out Joseph's disquiet with assuming uniformity between present and past rates of change. But when the Anthropocene started is unclear, with at least nine possible origin points in the last 50,000 years.[134] These range from global extinctions of megafauna during the last ice age, the expansion of farming 8,000 years ago, and the start of the Industrial Revolution in the eighteenth century. Finding a worldwide marker that geologists would accept, comparable to the Brunhes-Matuyama magnetic reversal (Figure 7.3), brings the start of the Anthropocene even closer in time. Some favour the detonation of the atomic bomb in 1945, others the exponential rise of plastic in shallow marine cores in the same timeframe.[135] Stratigraphically, as some see it, we have moved from the Pleistocene to the Plasticene.[136]

After Prehistory Comes Deep History

None of these histories was possible without the space made available by the time revolution and the need to fill it with content. But universal and big histories aside, the time revolutionaries would have gained most satisfaction from their legacy of deep history.[137] Deep human history, in its broadest sense, is Lubbock's prehistory and Evans's 'unwritten history'.[138] But the term prehistory has now served its purpose and stands awkwardly in the way of appreciating that *everything* is history: no pre-, proto- or post- is needed.[139] Their legacy of deep history is told through objects, the hard evidence of cats' tongues and flint fists, not just the archives of clay tablets, paper, and, more recently, magnetic tape and binary code. The narrative arc spans the history of the human brain, which evolved as powerful emotions were generated at social gatherings of both kin and strangers and which extended to the artificial objects and natural things without which such basic humanity could not exist.[140] The remarkable expansion of the brain over the last two million years was driven by our social lives, aided but also directed by our intimate involvement with the handheld, handmade object. This triangular tangle of brains, bodies, and objects co-evolved in ways we are only just beginning to appreciate. Other animals from chimpanzees to New Caledonian crows and warty pigs also make and use tools as well as being highly social. But none of them matches our unthinking ability to charge objects with emotion and meaning and to let them, on occasion, dictate how we should act. The time revolutionaries are prime examples, with their collections of coins, antiquities, geological fossils, beetles, and books. Hence Joseph's description of Jacques as an antiquary distinguished by an indefatigable zeal and perseverance. Deep history, as the time revolutionaries showed and their revolution established, is the story of our ages-long love affair with things.

Currently we are in a moment when history of all shapes and sizes is being written with objects. An archaeological biography of John Lubbock draws on the things he collected during his life.[141] A family history is told through antique Japanese netsuke, weaving these tiny carvings into the dramatic events of the twentieth century and the personal journey of the author.[142] A well-chosen hundred objects in the hands of an inspired narrator can tell the history of the world,[143] while the history of globalization is recounted through the fortunes of an empire of things.[144] We learn from these histories that objects, like people, acquire emotionally plangent biographies during their lives as they are skilfully made and used, lovingly repaired, wisely traded, and jealously guarded before being callously thrown away. As a result, the hard evidence for deep history comes to us well thumbed. And ever since the time revolution archaeologists have been answering John Evans's question 'How do we read the fingerprints?'[145] Some, like John, favour even more data and description. Others prefer to cloak the evidence with Toynbee's vivid sense of contemporary problems and dive into the theoretical questions about humanity that history always raises.

The result is a deep history that the time revolutionaries made possible. Simultaneously, they created the space needed to appreciate that history is tangible and that the rewards of zeal and perseverance are to be found in the common story of humanity, contained in the smallest and most insignificant of things.

Notes

1. Wallace 1898.
2. At the start of his long chapter on phrenology, Wallace quoted Caliban's line from Shakespeare's *Tempest*, 'All be turned to barnacles, or to apes with foreheads villainous low', to bolster his claims, which today are regarded as specious. The line is reminiscent of the developmental drawings by Charles Bennett which we saw in Chapter 5.
3. The Scottish phrenologist George Combe (1788–1858) was an admirer, for phrenological reasons, of George Eliot's large head, in which he saw the qualities of 'adhesiveness' and 'firmness' (Hughes 2017: Kindle p. 2395), while the author and politician Samuel Laing (1810–97) wrote, 'It is hard to say why, but as a matter of fact a weak chin generally denotes a weak, and a strong chin a strong race or individual' (1895: 395).
4. The X-Club was represented by Hooker, Lubbock, Spottiswoode, and Huxley (Moore 1982).
5. Wallace 1898 ch. XII.
6. Wallace 1898: 134.
7. Wallace had a longstanding interest in the preglacial settlement of America. Letter to B. Harrison, 20 January 1888 (Harrison 1928: 130).
8. Evans 1898.
9. Meltzer 2015.
10. Meltzer 2015: Kindle p. 5539.
11. Evans 1897.
12. See Chapter 6.
13. See Goodrum 2009 for an overview.
14. A magnificent triptych of portraits, Darwin flanked by Huxley and Lyell, hangs in the Nineteenth-Century Gallery.

15. Van Riper 1993: 215.
16. De Mortillet 1897.
17. Richard 2002.
18. Richard 2002: 181.
19. See Chapter 4.
20. Richard 2002: 181.
21. See Chapter 4.
22. https://openplaques.org/plaques/5860, accessed 3 August 2020.
23. http://falconermuseum.co.uk/, accessed 3 August 2020.
24. MacGregor 2008b.
25. Joan Evans (1964: 29) reveals that her father never catalogued any part of his collection of antiquities.
26. See Chapter 1.
27. Graham Richards's seminar on Lubbock at the Royal Institution, London, 26 June 1989.
28. Lubbock and Henslow both spoke after Huxley had answered Bishop Wilberforce (Owen 2013: 22). See Chapter 5.
29. http://www.abbeville.fr/loisirs/musee-boucher-de-perthes.html, accessed 3 August 2020.
30. M. F. Aufrère 2012.
31. The Ancient Monuments Protection Act was passed, after many difficulties, in 1882. Pitt-Rivers was the first Inspector of Ancient Monuments, appointed in 1883 (Owen 2013).
32. *SALON* 388, 5 June 2017.
33. A full list can be found in G. A. Prestwich 1899: 433.
34. Prestwich 1886, 1888.
35. G. A. Prestwich 1899. It is suggested (Boylan 1979: 177) that Grace was working on a biography of Uncle Hugh which she shelved to complete her husband's memorial. Her hand is evident in Charles Murchison's (1868) earlier biographical sketch and the letters used in her memoir by her sister Louisa (Milne 1901).
36. Geikie in G. A. Prestwich 1899: 420.
37. Geikie in G. A. Prestwich 1899: 420.
38. Hutchinson 1914. They played golf together as Hutchinson recalls (1914: 291).
39. Owen 2000 a & b, 2013; Thompson 2009; Clark 2014; Patton 2016.
40. The tradition continues. Arthur Evans has a chapter in *The Great Archaeologists* (Murray 1999); his father does not.
41. MacGregor 2008b. The volume arose from the well-funded Sir John Evans Centenary Project led by the Ashmolean Museum and supported by its royal patron Queen Margrethe II of Denmark, herself an archaeologist. Sherratt (2002) provides an example of Evans's gifts as a fundraiser for overseas scientific expeditions.
42. Patton 2016: Kindle p. 249.

43. Joan Evans added to her father's large collection of post-medieval finger rings, mostly posy rings, and donated them to the V&A in London (MacGregor 2008a: 147).

44. Evans 1964: 22. Writing after her mother died, Joan Evans recalled how Maria 'was determined that my existence should not spoil her life' (1964: 24).

45. Evans 1949: 124.

46. Evans 1943: 352.

47. MacGregor 2008c: 8.

48. *SALON* 390, 18 July 2017. Joan Evans recalled visiting her childhood home and finding it a shell (1964: 39).

49. https://historicengland.org.uk/listing/the-list/list-entry/1239636, accessed 3 August 2020.

50. Equivalent to £54 million and £436 million in 2018.

51. Lubbock 1904.

52. Evans 1955.

53. http://hamelinbrands.co.uk/john-dickinson-history/ 16.8.2016, accessed 3 August 2020.

54. http://johndickinsonwi.com/Products.aspx?CatId=11&SubcatId=58, accessed 3 August 2020.

55. Evans 1943: 326. It would be nice to think it was the 1859 vintage. If so, when last available at auction in 2017, a bottle fetched on average £5,361.

56. Compare G. A. Prestwich 1899: 399 with Dickens 1839 [1978]: 859.

57. Milne 1901 in G.A.Prestwich 1901.

58. The window is by Morris & Co., using one of Burne-Jones's designs. There is nothing geological about the imagery. The central angelic figure of Creation, above Joseph's name, is holding a chaotic globe representing the first day of creation.

59. *SALON* 388, 20 June 2017.

60. Owen 2013: 131.

61. Prestwich 1895.

62. Prestwich 1895: 1.

63. Prestwich 1895: 8.

64. Geikie 1894: 479.

65. Geikie 1894: 671. The thickness of loess in Eastern Europe and Russia reflects not only prevailing winds during the Pleistocene but also the continental conditions governing climate.

66. Evans 1897: 11.

67. Lubbock 1882: 28.

68. De Mortillet 1897: 321.

69. Penck and Brückner 1909.

70. The figure was calculated by Penck and Brückner.

71. Imbrie and Imbrie 1979: 123.

72. Imbrie and Imbrie 1979: 126.
73. Imbrie and Imbrie 1979: 140. The research involved oceanographers from Britain and the USA and included John Imbrie.
74. Bernard Brunhes in 1906 and twenty years later Motonori Matuyama, after whom these epochs of normal (as in the present day) and reversed polarity are named (Imbrie and Imbrie 1979: 147).
75. Imbrie and Imbrie 1979. Milankovitch's work was published during the 1930s.
76. Bridgland (1994) tied the Thames terraces into the deep-sea core chronology. Antoine (1990) previously did the same for the Somme, where ten distinct terraces are recognized during the last million years (Antoine et al. 2011; Locht et al. 2013; Hérisson et al. forthcoming).
77. Bridgland (2014) provides a summary of Lubbock's insights. See Chapter 5.
78. Jean-Luc Locht, pers comm. The fine sediments in these pits date to stage 11 in the marine record. For the detailed geology, see Antoine et al. (2011: fig. 14) and Locht et al. (2013).
79. Ashton 2017.
80. Bahain et al. 2016: 365–6; Moncel et al. 2016.
81. Leakey, Evernden, and Curtis 1961.
82. Evans 1897: 15.
83. Now classified as *Homo erectus*.
84. See Gamble 1993 for the changing locations of the cradle of humanity.
85. Harmand et al. 2015.
86. McNabb 2012: ch. 9.
87. Harrison 1928: 79 recalls when the two men met and their friendship began. G. A. Prestwich. 1899: 304, Prestwich, Letter to Evans, 10 October 1879. See Prestwich (1895: ch. III) for these curious plateau implements.
88. Van Riper 1993: 212.
89. Prestwich 1895: 57–8.
90. Prestwich 1895: 59 and 77.
91. McNabb 2012. Harrison 1928 has several references to Joseph and Grace Prestwich's meetings with Boucher de Perthes.
92. Evans had a frosty correspondence with Harrison about the non-implements that he and Joseph regard as preglacial in age (Harrison 1928). See White (2017: 214–16) for the sides in the debate that pitted Evans against Prestwich.
93. Spencer 1990.
94. Owen 2013: 131.
95. Lubbock 1913: 337; Owen 2013: 131.
96. Dart (1925) named the juvenile skull *Australopithecus africanus*, the 'southern ape of Africa'.
97. Many names have been put forward as the hoaxer(s). For example, Spencer (1990) fingers the august figure of the human palaeontologist Sir Arthur Keith but without convincing other Piltdown specialists. The tools of the forger were

found in the Natural History Museum in 1996 and point to one of its own scientists, Martin Hinton (Gardiner 2003). Until the next revelation the contents of Hinton's trunk, for the moment, seem conclusive. Hinton was a member of Harrison's Ightham eolith group. He had opportunity and the motive of professional disillusionment. Charles Dawson the 'excavator' is also heavily implicated, but conclusive proof is lacking (Ashton 2017).

98. See Chapter 4.
99. Meltzer 2009: 86–7. The photograph taken on 4 September 1927 (Meltzer 2009: fig. 16) has Schwachheim pointing to the projectile point that was found in late August and left in the ground, 'guarded day and night, until a panel of experts could arrive'.
100. Meltzer 2009: 292 uses the uncalibrated age of 10,490. I have calibrated this using the CalPal programme available online. The revised date creeps into the Younger Dryas, a short-lived return to glacial conditions before the resumption of warming that initiates the Holocene 11,500 years ago.
101. Adovasio and Page 2002.
102. Meltzer 2009: 199.
103. Gamble 2013.
104. Acheulean bifaces have been found with fossils of *Homo ergaster*, *Homo erectus*, *Homo heidelbergensis*, and an early representative of *Homo sapiens* (Gamble 2013; Gamble et al. 2014), covering a time span from 1.8 million years to 200,000 years ago.
105. Roe 1968a, 1968b. His intermediary group showed no clear preference for either form, while Evans's third group was based on flakes rather than bifacial hand-axes. Roe later indicated that his intermediary group could 'legitimately be ignored' (1981: 154).
106. Gardner 1965: 41–2; Carroll 1865.
107. Dickens 1865: 1.
108. Evans 1943: 106.
109. Evans 2012: 296.
110. Toynbee 1884: 5, published posthumously. See https://archive.org/details/industrialrevol00toyngoog/page/n20, accessed 3 August 2020.
111. See Wells (1922) for the potted, short history of the world. His much longer *Outline of History* appeared in 1920 after it had been serialized the year before. McNabb (2015, forthcoming) examines Wells's interest in human evolution.
112. See Chapter 5.
113. Wells 1921: 287.
114. These same passions were on show before the Great War in the bestselling *La Guerre de feu* (Quest for Fire) written by J. H. Rosny-Aîné (1866–1940) in 1911. Alpha male Neanderthals fight over their women using 'brutal strength, tireless and without pity'. And to the victor went the woman who would kneel to her master, carry the animals he killed back to the cave, and, if disobedient, she

would be put to death. Ominously from a Darwinian perspective, 'what seemed like time without end stretched out before them'.

115. Stevenson's novel appeared in 1886.

116. See Moser (1998) for the iconography of human origins, and Horrall (2017) for the invention of the caveman from E. T. Leeds's 'Prehistoric Peeps' in *Punch* in 1893 to George Robey dressed as a humorous caveman in 1902, to *The Flintstones* in the 1960s, and up to date with Aardman's *Early Man* (2018).

117. Griset was a popular illustrator of fairy tales, to which he added a grotesque and. frankly, terrifying vision. His illustrations for James Greenwood's *Legends of Savage Life* (1867) (see https://archive.org/details/legendssavageli00grisgoog, accessed 3 August 2020) would have brought him to Lubbock's notice. Victorians found these comic sketches of 'primitive' races highly amusing, and by comparison his scenes of Palaeolithic life for Lubbock were restrained.

118. Murray 2009.

119. Owen 2013: 110.

120. The pictures can be viewed at https://www.latrobe.edu.au/humanities/ research/research-projects/past-projects/lubbock-gallery, accessed 3 August 2020. They are a sober version of the later 'Prehistoric Peeps' discussed by Horrall (2013).

121. Woodward 1893: 246.

122. Issac Asimov's description of Timmie in *The Ugly Little Boy* (Asimov 1958) is a further instance of popular rehabilitation. Timmie can be taught to speak, be civilized, and call his nurse Mother.

123. These novels drew on the scientific reassessment of Neanderthal skeletons and skulls that had been deliberately misrepresented in the early years of the century, particularly the 1908 find of a complete skeleton from La Chapelle-aux-Saints, France (Stringer and Gamble 1993; Papagianni and Morse 2013).

124. It was this question, rather than religion, that divided opinion (Patton 2016: Kindle p. 66. See Chapter 5.

125. Montagu 1972: 238. See Gamble 2007: ch. 2 for discussion.

126. Greene 1999. Harari (2011) uses three revolutions—cognitive at 70,000, agri-cultural at 12,000 and scientific starting 500 years ago—to structure his universal history. Childe was also following de Mortillet (1897), who argued for the Neolithic as the hinge in human history.

127. This phrase appeared on the dust jacket of one of his last books *The Prehistory of European Society*, 1958, the year he committed suicide in Australia, but echoes the sentiments of his earlier works in the 1940s written during the Second World War.

128. Mellars 1973; Klein 1973. For updates, see Gamble (2007); Shea (2011); and Cook (2013).

129. McBrearty and Brooks 2000.

130. Finlayson et al. 2012; Zilhao 2007.

131. Hockett and Ascher 1964. Montagu (1965: 15) places the human revolution with the oldest stone tools.
132. See Knipe (1905: 190), with an accompanying illustration of Early Palaeolithic man chipping flints while a woman tends a baby.
133. Christian 2004.
134. Lewis and Maslin 2015.
135. Brandon et al. 2019; Steffen et al. 2011.
136. http://www.earthdecks.net/plasticene/, accessed 3 August 2020.
137. Shryock and Smail 2011.
138. Evans 1882.
139. Gamble 2014; Fontijn 2016; Smail and Shryock 2013.
140. Smail 2008; Shryock and Smail 2011; Gamble, Gowlett, and Dunbar 2014.
141. Owen 2013.
142. De Waal 2010.
143. N. MacGregor 2010.
144. Trentmann 2016.
145. Evans 1882a, 1882b.

8

Afterword

The Stone That Shattered the Barrier of Time

I became interested in the time revolution when I did a short talk on *Strange Encounters* for BBC Radio 3.[1] What could be stranger, I thought, than two geologically minded English businessmen encountering a rough stone axe in a French gravel pit made by a remote, but invisible ancestor who knew intimately what a woolly mammoth smelt like? I had always known about Prestwich and Evans and been taught they were part of the *annus mirabilis* of 1859.[2] Off they went to Abbeville to check out the madcap claims of Boucher de Perthes, and, unexpectedly, the story goes, they came back with the proof that dealt a body blow to a biblical timescale. Then the histories of archaeology move rapidly through that year to Darwin's bombshell of a book and its impetus to unearthing evidence for human origins: Neanderthal skulls, validating cave art, and amassing enormous amounts of stone tools from around the world.

I trawled through the literature and realized there was more to the story. For a start, no one knew when exactly in 1859 their dramatic discovery took place. Sometime in spring was as good as it got. Then, there was little mention of how they made their case and what obstacles they had to face. It struck me as curious that a revolution in time was blithely unconcerned with the age of what they found. Chronology does matter, especially to an archaeologist like me.

Then, there was the question of expertise. John Evans is heralded, rightly, as the great authority on Palaeolithic artefacts. But in 1859 he was a papermaking numismatist with a geological hobby. He hadn't seen anything like Boucher de Perthes's stone axes before the April trip. It was a good job he was a quick learner. But why did the histories I was reading dwell only on the geological and archaeological skills of Prestwich, Evans, and Lubbock and not on their hinterland in commerce, manufacturing, and industrial design, which gave them a distinctive world view? Why maintain the

Victorian fiction of separate spheres to keep antiquarian and geological pursuits apart from business acumen and political events? Scientific discoveries never take place in isolation, and 1859 was no different as the time revolutionaries pursued the 'truth' of the matter. The decade from 1859 to 1872 was as turbulent as any other. It saw, among many conflicts, the Indian Rebellion, campaigns in the Italian Risorgimento, the American Civil War, and the Franco-Prussian War. Attitudes to colonialism, race, and slavery and the boundaries of nation states were shaped by these historical earthquakes and collapsing calderas. Human antiquity was swept up and shaped by these forces.

The more I looked into the *annus mirabilis*, the more I realized that these men were driven by the principles in Samuel Smiles's *Self-Help*, published on the same day as *The Origin*, where zeal and perseverance are held up as virtues, rather than by Darwin's struggle for survival, which is where their contribution is usually parked. It was simply enough that Darwin approved of what they were doing. All of them could recognize bad geology when they saw it. John and Joseph weren't out to bring down religion or shock Victorian morality by inviting a distant, primeval relative into the drawing room to meet the vicar for tea. They just wanted to correct a misconception that offended their passion for geological exactitude:; humans *had* to have lived alongside mammoths. They were the men to prove it. And they set out to turn a discovery into a scientific experiment repeated across Europe and the globe.

But what really struck me was the absence of the worked stone that they diligently photographed, described in detail as it poked out of its gravel bed like a cat's tongue, and paraded before the learned societies of London. Why was such a well-documented object so mysteriously absent from all the textbooks and histories? And why didn't they publish a drawing of it? Most intriguingly, could the missing stone be found?

It was not the first bit of evidence to have gone missing. Their photographs, which so impressed Lyell, Murchison and all who saw them, had also disappeared. Joan Evans recalled seeing them as a girl, but by the 1940s they had vanished from the family archive, and the copies given to the Society of Antiquaries and Royal Society were lost.[3] All that remained was a blob of glue in Prestwich's manuscript (Figure 3.6).

The originals were found in 1978 by Micheline Agache-Lecat in the Amiens library.[4] Charles Pinsard had left the city his descriptions of its buildings, street by street, in seventy-two large volumes handsomely bound in red leather. The two photos are in volume 43, firmly stuck to the page surrounded by his copperplate writing describing the day's events.[5] A few pages buried in an immense architectural history of an ancient city.

And then forty years later, and long after I had given up hope that other copies might be found, Prestwich's copies were rediscovered in London. Caroline Lam, an archivist at the Geological Society, recognized their importance in 2018 while she was researching Joseph's involvement in Brixham Cave. They had been bound into a volume of his offprints, among them his 1860 and 1864 papers on the Somme in the *Philosophical Transactions*, together with watercolour copies of the stone tools drawn by his sister Civil for the Royal Society. The two photographs are not glued down, and the thin paper they were printed on has aged remarkably well. They are exactly the same size as the Pinsard prints in Amiens.[6]

The photographs are all that survives of Fréville's pit. Houses have since spread along the Cagny Road. But in one unbuilt corner the city of Amiens has created an evocative Jardin d'Archéologique. You can walk along a timeline back to 400,000 years ago to find a vertiginous section of gravels and sands like those which greeted John and Joseph on that Wednesday afternoon in 1859 (Figure 8.1).[7] Climbing the observation tower, you look out over the river terraces and the houses which now fill the deep history pits of Saint-Acheul. This is where the time revolution planted its banner.

But did the loss of the discovery flint really matter? The archaeologist Chris Evans points out that when it comes to answering archaeology's big questions, such as the antiquity of humans, this wasn't done by a single excavated find but by the thousands of similar artefacts recovered subsequently. This goes to show that archaeological excavation is a repeatable experiment and not a one-off discovery.[8] Archaeologists don't destroy the past by digging it up. Rather, they confirm its essential outline in time and geographical space. And this is because of the numbers involved, something Joseph and John and before them Frere, Rigollot, and Boucher de Perthes latched onto from the very start: the quantities of stone tools used by the gravel diggers to fill the ruts in the roads of Hoxne and Abbeville. Finding one missing flint wouldn't change the course of deep history. But if it could be found, I would have my own experiment in resolving the scales of deep history and archaeological history, the present and the past, the facts and the theory, the power of time to shape, and its potency to unite through the tangible object.

Figure 8.1 The Jardin d'Archéologique at Amiens, very close to Fréville's pit. Prestwich and Evans would have been interested in level 1, as shown on the key, dated to 430,000 years ago and containing the *terrassiers' langues de chat* in a flint-rich gravel within a sandy matrix.

Source: Author's photo.

Could the Flint Be Found?

With the hundred and fiftieth anniversary due in 2009, I set out the year before to see if the iconic stone for the time revolution still existed. I started with the likely places, the Ashmolean Museum because of its Evans connections, the Society of Antiquaries, which has one of Frere's Hoxne hand-axes, and the British Museum, which has the other on display in its Enlightenment Gallery alongside the people's axe from Black Mary's. I drew a blank with all three.

That left the Natural History Museum. Its collections were once part of the British Museum, but in 1881 the doors opened on this new museum for all things living, dead, and evolutionary. I contacted one of the curators I knew, Robert Kruszynski, who had worked for many years in the Department of Palaeontology.[9] He knew his collections inside-out and emailed me back saying that Grace Prestwich had sent Joseph's collections there after he died in 1896. He promised to go and have a look in the catalogues and the drawers of stone tools.

He replied the next day.[10] He had traced in the catalogues some flint tools that were found in 1859. When they arrived in 1896, fifty-one of them from

Saint-Acheul had been lumped under a single accession number, the prosaic E373. Then, in 1953, the Prestwich collection had been recatalogued, and each of these implements given a separate number. Even better, one of the flints had a label on it with some interesting words. He had taken photographs. I was extremely excited as I clicked on the attachments.

And there it was. The first photo showed a small hand-axe, too big for its cardboard tray. It was clearly a work in progress, roughly shaped and triangular (Figure 8.2). One of Evans's pointed types. And Rob had included a second picture of two labels stuck to it. An eye-catching yellow adhesive dot with the catalogue number it was given in 1953, E5109. And above the dot was a small rectangular strip of white paper held steadfastly in place by Victorian glue that read, 'St Acheul Amiens 11ft from surface April 27-59'. This was it. The depth of eleven feet corresponded to the photograph taken on the day (Figure 2.10) and to Joseph's scribbles in his notebook kept in the Geological Society Library. And then in a different, smaller hand someone had added 'Present when found J.P.' None of the other flint tools Prestwich brought back has such a label. I like to think that Civil, his sister, wrote and glued the original label, while either Joseph or Grace added the detail later about being present. They wanted posterity to know he was a witness to history.

A week later I went up to London to see for myself, clutching a copy of the photograph of the flint sticking out of the gravels. Would it match? To be true to Joseph and John, I wanted witnesses. So, with Rob and his colleagues Chris Stringer and Mark Lewis, we compared the artefact and the photo separated from each other by almost 150 years. The match was perfect.[11] Along with the discovery flint Rob had found the implements Prestwich illustrated in his Royal Society paper. We placed each artefact on top of its published drawing. All were present and correct.

And there was more. Rob had also unearthed a small oval hand-axe, given to Joseph by Jacques in April 1859. A larger label is stuck to it with a description in Jacques's distinctive, flowing script. It's difficult to read (Figure 8.3) but he states that this flint was found, as Prestwich wanted, near an elephant tooth. This was the proof of antiquity by association. People had once lived alongside mammoths. The two flints, pointed and oval, from Amiens and Abbeville respectively, are the foundation stones of deep human history.

When I handled the discovery flint that validated Boucher de Perthes's experiment in deep history, I could understand straight away Joseph's scribbled note in his Royal Society manuscript: this was not a piece that

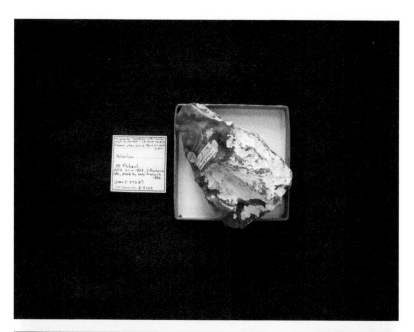

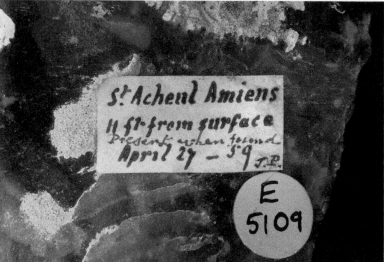

Figure 8.2 The discovery flint in the Natural History Museum, London. Note the feathery flaking below the label.

Source: © Natural History Museum, London/Phil Crabb.

Figure 8.3 An oval flint implement given to Joseph by Jacques in April 1859. The label, deciphered by Pierre Schreve, reads,

> Abbeville Porte Mercadé
> Banc de Menchecourt
> 6 mètres de profondeur
> 50 cent[imetres] en dessous du
> niveau de l'eau, au
> lieu visité par Mr J
> Prestwich avril 1859.
> J Boucher de Perthes
> non loin du lieu où
> était la dent d'un jeune elephant.

Source: © Natural History Museum, London/Robert Kruszynski.

would have been selected on its looks. Small and irregular, with a chalky surface that comes off on your hands, it was overshadowed by the finely fashioned examples they did illustrate. One further thing caught my eye. Civil must have stuck its label on as soon as they got back to London. It was probably done for the Flint soirée, and so before John had his heart-stopping moment in front of a display case in Somerset House. Why? Because its positioning, which makes the pointy end face downwards, is the same as the watercolour sketches Civil prepared for the engravers (Figure 3.5). Everything changed for the final publication, when the direction was reversed to follow the orientation used by Frere for the larger of his two hand-axes in 1800. But by then the label on the discovery flint was permanent, and soon the implement would be overlooked. Instead, the flint axes they drew jump out from the page with a simple message: history points upwards, as Lubbock always claimed. This dynamic layout enhances the evidence for intelligent design, as demonstrated by symmetry and finer flaking. Only with such pieces drawn in this manner could they convince a sceptical reader.

But I also wonder if they had second thoughts about the authenticity of the discovery flint. Some of the flake scars are fresh and layered like feathers (Figure 8.2), surprising for a flint artefact found in such chaotic gravels. Surely its edges would be abraded, smoothed by rolling in the river, as is the case with Jacques's flint (Figure 8.3)? And there's more. A prominent bulb of percussion at its thick base looks as if it was hit by a metal object. That could have been an innocent blow from a *terrassier*'s shovel or, more sinisterly, from a forger's knapping hammer. Did Evans have doubts about it as his expertise in flint knapping grew during 1859 and his nose for spotting a forgery became keener? If so, then it might explain why it was never drawn, quietly forgotten about, and kept out of sight, safely buried in Joseph's collection. That might also explain why he didn't include drawings of the photographed sections in Fréville's pit, where it was found. Having trumpeted the discovery on their return to London, they then hid the physical evidence, providing only a written description of the discovery that couldn't be challenged. In the absence of written evidence to explain their actions, all we have are questions about the object itself. All very deep history. Perhaps it's a case, as John Lubbock would say, of 'what we see depends mainly on what we look for'.[12]

But enough! I can hear Prestwich's booming voice admonishing me, 'Stick to the published facts and don't indulge in unsupported conjecture. Evidence trumps theory every time. We were the expert witnesses. I was present'.

No matter. Whatever its age and whoever made it, it's still the flint discovered on Wednesday afternoon, 27 April 1859. The stone that shattered the barrier of time and ushered in the revolution of deep human history.

Notes

1. Produced by Roland Pease and broadcast April 2007.
2. Daniel 1964: 45.
3. Evans 1949: 122 n. 25.
4. Cohen 1998. Jean-Pierre Fagnart, email to the author, 16 September 2008.
5. MS 1370E *Recueil de notes relatives à l'histoire des rues d'Amiens* in the Bibliothèques d'Amiens Métropole. There are additional pages of his own fine drawings of stone tools from Saint-Acheul.
6. *Joseph Prestwich Tracts on the Antiquity of Man 1859–1875* LDGSL/800a Geological Society. Caroline Lam provides a full description of this remarkable volume in her catalogue entry. It was shelved in the rare books collection but has now been transferred to the archive.
7. The section was dug by Victor Commont (1866–1918), a leading light in the next generation of French Palaeolithic archaeologists. The pit was owned by Bultel and Tellier.
8. Evans 2012: 296.
9. Gamble and Kruszynski 2009.
10. 22 July 2008.
11. See Chapter 2. Original image: *Antiquity*, doi.org/10.1017/S0003598X00098574.
12. Lubbock 1892: 4.

Glossary of Terms Used in 1859, and Their Modern Counterparts

Archaeology and Artefacts

Biface French archaeologists' current term for Boucher de Perthes's axes. A biface is knapped on both surfaces (bifacial) to form the pointed and oval forms which experiments show are efficient tools for butchering large animals. The term is used also by English-speaking archaeologists.

Blade Distinguished from a flint **flake** by being at least twice as long as it is wide. The blank from which **Upper Palaeolithic** tools were fashioned.

Celt Pronounced with a soft 'C' as 'Sɛlt'. No longer used. From the Latin for 'chisel' or 'adze' and sometimes used to describe Boucher de Perthes's axes. 'Celts in the Drift' was used as a byline in 1859 publications, but does not refer to the ancient **Celts**.

Coup de poing 'Punch', a former term for ice age **hand-axes**. It suggests they were used with a downward stabbing motion.

Eoliths 'Dawn stones'. Claimed to be older than the **hand-axes** found on the Somme and thought by de Mortillet and Prestwich to date to preglacial times. Most of them are natural stones, and eolith is no longer used (see Chapter 7).

Flake A purposefully struck segment from a stone nodule. Flint flakes come in all shapes and sizes and can be worked further into a variety of tools for cutting, scraping, and as projectile points. **Bifaces** can also be made on large flakes.

Hache ébauchée Rough stone **hand-axe**, as described by Boucher de Perthes and Rigollot (see Chapter 2).

Hache antédiluvienne Ice age stone **hand-axe**, as described by Boucher de Perthes (see Chapter 2).

Hand-axe English equivalent of **biface** and widely used to describe the Lower Palaeolithic axes that Boucher de Perthes first described. The name describes the unhafted nature of these handheld tools.

Knapping The intentional flaking of stone to produce either **flakes** or to fashion **bifaces** from nodules of raw material. **Palaeolithic** knapping involved the use of two types of hammer; hard (another stone) for roughing out the shape and soft (wood, bone, or antler) for fine finishing.

Langues de chat 'Cat's tongues'; the name given to the ice age **hand-axes** by the *terrassiers* of Abbeville and Amiens (see Chapter 2).

Neolithic 'New Stone Age'. First used by Lubbock in 1865 and identified through polished stone axes, farming' and megalithic monuments such as Stonehenge, Avebury, and Carnac (see Chapter 6).

Nodule The raw material used by Palaeolithic stone knappers, found in beaches, river gravels, and occasionally as outcrops. In northern Europe flint nodules are common and have good flaking properties. Many other rock types were used.

Palaeolithic 'Old Stone Age'. First used by Lubbock in 1865. All Boucher de Perthes's and Evans's finds of unpolished stone axes were placed in it. Hunting, gathering, and fishing formed the mainstay of the domestic economy. Subdivided by de Mortillet into Lower, Middle, and Upper Palaeolithic according to the types of stone tools and the presence of art (see Chapter 6).

Geology and Chronology

Antediluvian Before Noah's flood or any earlier cataclysmic flood known to geologists.

Diluvium/Catastrophist theory Boucher de Perthes argued in 1847 that the ice age animals and the makers of the stone tools humans found in the Somme had been destroyed by a cataclysmic flood. As a result, humans had to be created again. The idea of evolutionary continuity between ancient and modern humanity did not enter into his scheme. He regarded Noah's flood as only the most recent deluge. Prestwich did not share these archaic geological views.

Drift, The A surface geology of soft, rather than hard rock sediments, predominantly sands and gravels laid down as a result of river or ice action, and also incorporating fine-grained, windblown loess. The Drift formed regional geologies and provided evidence for multiple flood events. It was in these sediments that bones of extinct animals and stone implements were found. In 1839, Drift replaced, for English geologists like Prestwich and Falconer, the older term **Diluvium** inspired by the global biblical flood. Drift deposits are now termed **Pleistocene** or **Holocene** depending on their age (see Chapters 1 and 2).

Mosaic chronology The adjective means 'relating to Moses'. The chronology adds up the ages of the Prophets and Elders in Genesis. One calculation by Archbishop Ussher of Armagh in 1650 gave the age of the Earth as beginning on 23 October 4004 BC, 6,000 years ago (see Chapter 4, Table 4.1).

Postdiluvian After Noah's flood.

Principle of superposition Stratigraphy is possible because, as Charles Babbage put it, 'the order of succession of strata indicates the order of their antiquity, the lowest being always the oldest'.

River Terrace As rivers cut down they leave behind remnants of earlier levels where they once flowed. The effect is clearest along the sides of major rivers where a flight, or staircase, of terraces can often be found. The higher the terrace, the older it is in the sequence. The Somme's flight of terraces was scooped out of the chalk bedrock as the river lowered the valley floor. The benches left behind contained sands and gravels deposited by the river when it flowed at that height. It was here that the *terrassiers* dug for ship and railway ballast and road and building material.

Stratigraphy The study of the Earth's rocks as distinct strata determined by their chemical and physical composition and the animals and plants they contain.

Terrassiers French quarrymen who were proficient at finding hand-axes in the gravel pits of the Somme. Key witnesses and enablers of the time revolution.

Uniformitariansim The geologist's bedrock principle, which states that processes which operate in the present day, such as mountain building or the erosive power of glaciers, also operated in the past and produced the same results. Opinions differed about the rates of those changes, which may not have been the same in the past.

Time	Today	1859
↑	Holocene ('Entirely new') epoch introduced by Lyell in 1839 but not commonly used in 1859. Before the Anthropocene was named in the early 2000s, it was the epoch of the present day	Post-Glacial & Recent came after the single ice age of 1859, when the climate warmed and the cold-adapted fauna of woolly mammoth and rhino disappeared.
	Pleistocene ('Newest') epoch introduced by Lyell in 1839. In 1859 the Drift, proposed by Murchison also in 1839 to replace Diluvium with its connotations of Noah's flood, was more commonly used.	Post-Pliocene ice age. Evidence for Alpine and Scandinavian glaciation began with Agassiz in 1840. In 1859 Lyell and Prestwich believed in a single ice age preceded by the Preglacial and followed by the Post-Glacial. The work of Croll and Geikie in the 1860s demonstrated multiple cold glacial phases when the ice advanced, interspersed with evidence for warmer, interglacial phases comparable to the temperate climate enjoyed by the nineteenth century.
	Pliocene ('Newer') epoch introduced by Lyell in 1839	Preglacial the period before the single ice age of Lyell and Prestwich. Finding stone tools such as eoliths in these deposits would make them very old indeed

Human Evolution and Anthropology

Anthropology The study of humans in all their dimensions, physical, cultural, archaeological, and linguistic. It replaced **ethnology** as the name for the field in the late nineteenth century.

Celts and Celtic Pronounced with a hard 'C'. Boucher de Perthes made the stratigraphic distinction between a recent Celtic, and an older **antediluvian** past. Celts inhabited large parts of Europe. They were either subjugated by the Romans or lived beyond the empire, as in Wales. The existence of Celtic peoples has been questioned by some archaeologists.

Development drawings Widely used by scientists, numismatists such as Evans (see Chapter 3), and satirists like Charles Bennett (see Chapter 5) to either show or lampoon how one thing can become another through incremental changes over time. The process is guided by errors in copying that arise from imitation.

Ethnography In 1859 described the customs, habits, and beliefs of peoples. It provided the raw material for **ethnology**. In practice ethnographers focused on peoples outside the industrial cities and agrarian economies of the nineteenth century, as in the United States, and living under European Imperial rule. Museums of ethnographic objects, such as those collected by General Pitt-Rivers, were displayed according to ethnological ideas.

Ethnology In 1859 the science that dealt with classifying human populations into races by studying their origin, geographical distribution, and cultural and biological characteristics. Lubbock saw himself as an ethnologist.

Lamarckism As proposed in 1809, characteristics acquired during one lifetime are passed on to the next generation. This explanation of biological evolution was overtaken in 1859 by Darwin and Wallace's demonstration of descent with modification under **natural selection**. Today Lamarckism is the basis for understanding cultural rather than biological evolution. For example, Jacques Boucher de Perthes inherited his father's antiquities and added to them.

Monogenist Someone who believes that there was a single original ancestor from which all the varieties of living people are descended. Darwin and Wallace's natural selection with modification by descent provided a mechanism whereby geographical populations (races) evolved from a common ancestor.

Natural selection Darwin and Wallace's mechanism to explain how life changes and where the environment, physical and social, determines which members of a population survive to reproduce.

Polygenist Someone who believes that the varieties of living people are original, fixed from the beginning, and therefore evolution is unnecessary to explain differences between geographical populations (races).

Pre-Adamites The people who filled geological time before Adam's creation, as recorded in Genesis (see Chapters 1 and 4).

Transmutation Change from one form into another.

Learned Societies and Clubs

Académie des Sciences Founded in 1666, it became the leading learned society in France for the study of science. Its *Comptes Rendus*, 'Proceedings', published scientific discoveries from its base in Paris.

Anthropologicals A breakaway group from the Ethnological Society led by the racist James Hunt. His Anthropological Society existed from 1863 to 1871, when it was repurposed by Lubbock and others as the Anthropological Institute, which incorporated both learned societies.

Antiquary Term used to describe those like Evans and Lubbock with a curiosity about the past and its study through objects and texts. It was gradually replaced by **archaeologist**, which, by the beginning of the twentieth century, had become the common descriptor of professionals and amateurs alike.

Ethnologicals Numbered among these were Huxley, Tylor, Evans, and Lubbock and others with a progressive view of race and a track record in opposing and abolishing slavery.

Geological Society Founded in 1807 and today based in Burlington House, Piccadilly. Its motto is *Quicquid sub terra est*, the practical study of 'whatever is under the earth'.

Royal Society Founded in 1660, the leading scientific society in Britain. It is now based in Carlton House Terrace, London. Its motto is *Nullius in verba*, 'Take nobody's word for it.'

Société (Royale/Impériale) d'Émulation d'Abbeville Abbeville's Learned Society for the study and promotion of literature, science, and the arts. Founded during the French Revolution, it was presided over for many years by Jacques Boucher de Perthes. The SEA continues to thrive see https://www.societe-emulation-abbeville.com/, accessed 5 August 2020.

Société des Antiquaires de Picardie Founded in 1836 for the study of the art, history, and archaeology of Picardy. It is based at the Amiens Museum in rue de la République.

Society of Antiquaries of London Founded in 1707 for the 'ingenious and curious' and now based in Burlington House, Piccadilly. Its motto is *Non extinguetur*, which translates as '[the lamp (the knowledge)] will not go out'. This refers to the symbol of the Society.

X-Club An influential dining club of nine leading London scientists that first met in 1864 and was disbanded in 1893. Lubbock was a member of this Victorian version of a think tank (see Chapter 6). All of its members also belonged to the older *Athenaeum* club, which was a major force in promoting Victorian science.

References

Adhémar, J. A., 1842. *Révolutions de la mer*. Paris: Privately Published.

Adovasio, J. M. and J. Page, 2002. *The First Americans: In Pursuit of Archaeology's Greatest Mystery*. New York: Random House.

Agache, M., 1974. 'Une Lettre de Boucher de Perthes à Isidore Geoffroy Saint-Hilarie'. *Cahiers archéologiques de Picardie*, 1, 5–10.

Agassiz, L., 1840. *Études sur les glaciers*. Neuchâtel: Jent and Gassman.

Altholz, J. L., 1982. 'The Mind of Victorian Orthodoxy: Anglican Responses to 'Essays and Reviews,'1860–1864'. *Church History*, 51(2), 186–97.

Antoine, P., 1990. *Chronostratigraphie et environnement du Paléolithique du bassin de la Somme*. Lille: Publications du CERP 2.

Antoine, P., J.-J. Bahain, P. Auguste, J.-P. Fagnart, N. Limondiin-Lozouet, and J.-L. Locht, 2011. 'Quaternaire et préhistoire dans la vallée de la Somme: 150 ans d'histoire commune', in A. Hurel and N. Coye, eds., *Dans l'épaisseur du temps: Archéologues et géologues inventent la préhistoire*. Paris: Publications Scientifiques du Muséum national d'Histoire naturelle, 341–82.

Ashton, N., 2017. *Early Humans*. London: Harper Collins.

Asimov, I., 1958. 'The Ugly Little Boy', in I. Asimov, ed., *The Best Fiction of Isaac Asimov*. London: Grafton.

Aufrère, L., 1940. 'Figures des préhistoriens,1 Boucher de Perthes'. *Préhistoire*, 7, 1–134.

Aufrère, M.-F., 2012. 'Histoire de l'archéologie préhistorique comme patrimoine'. *Les Nouvelles de l'Archéologie*, 129.

Babbage, C., 1859–60. 'Observations on the Discovery in Various Localities of the Remains of Human Art Mixed with the Bones of Extinct Races of Animals'. *Proceedings of the Royal Society of London*, 10, 59–72.

Bahain, J.-J., N. Limondin-Lozouet, P. Antoine, and P. Voinchet, 2016. 'Réexamen du contexte géologique, chrono- et biostratigraphique du site de Moulin Quignon à Abbeville (Vallée de la Somme, France)'. *L'Anthropologie*, 120, 344–68.

Baines, T., 1866. 'On the Flint-Flakes in the Drift and the Manufacture of Stone Implements by the Australians'. *Geological and Natural History Repertory*, 13, 258–62.

Barton, R., 1998. '"Huxley, Lubbock, and Half a Dozen Others": Professionals and Gentlemen in the Formation of the X Club, 1851–1864'. *Isis*, 89, 410–44.

Barton, R., 2018. *The X Club: Power and Authority in Victorian Science*. Chicago: University of Chicago Press.

Bayly, C. A., 2004. *The Birth of the Modern World 1780–1914*. Oxford: Blackwell.

Beckwith, R., 1994. 'Essays and Reviews (1860): The Advance of Liberalism'. *Churchman*, 108, 1–10.

Bennett, C. H., 1872. *Character Sketches, Development Drawings, and Original Pictures of Wit and Humour*. London: Ward, Lock, and Tyler.

Berggren, W. A., 1998. 'The Cenozoic Era: Lyellian (Chrono)Stratigraphy and Nomenclatural Reform at the Millennium'. *Geological Society, London, Special Publications*, 143, 111–32.

Boucher de Perthes, J., 1842. *De l'éducation du pauvre: quelques mots sur celle du riche discours prononcé par le Président de la société royale d'Emulation d'Abbeville, dans la séance du 29 octobre 1841*, Abbeville: Briez.

Boucher de Perthes, J., 1847. *Antiquités celtiques et antédiluviennes. Mémoire sur l'industrie primitive et les arts à leur origine*. Volume 1. Paris: Treuttel & Wurtz.

Boucher de Perthes, J., 1857. *Antiquités celtiques et antédiluviennes. Mémoire sur l'industrie primitive et les arts à leur origine*. Volume 2. Paris: Treuttel & Wurtz.

Boucher de Perthes, J., 1860. *De la femme dans l'état social, de son travail et de sa rémunération: discours prononcé à la société impériale d'Emulation d'Abbeville, dans la séance du 3 novembre 1859*, Abbeville: Briez.

Boucher de Perthes, J., 1864a. *Antiquités celtiques et antédiluviennes. Mémoire sur l'industrie primitive et les arts à leur origine*. Volume 3. Paris: Treuttel & Wurtz.

Boucher de Perthes, J., 1864b. 'De l'homme antédiluvien et de ses œuvres', in J. Boucher de Perthes, ed., *Antiquités celtiques et antédiluviennes. Mémoire sur l'industrie primitive et les arts à leur origine*. Volume 3. Paris: Treuttel & Wurtz, 1–105.

Bowler, P., 2009. 'Darwinism, Creative Evolution, and Popular Narratives of "Life's Splendid Drama"'. *British Academy Review*, 14, 46–9.

Boylan, P. J., 1979. 'The Controversy of the Moulin-Quignon Jaw: The Role of Hugh Falconer', in L. J. Jordanova and R. S. Porter, eds., *Images of the Earth: Essays in the History of the Environmental Sciences*. Chalfont St. Giles: British Society for the History of Science, 171–99.

Boylan, P. J., 2008. 'Evans, Geology and Palaeontology', in A. Macgregor, ed., *Sir John Evans 1823–1908: Antiquity, Commerce and Natural Science in the Age of Darwin*. Oxford: Ashmolean, 52–67.

Bradshaw's, 1853 [2016]. *Continental Railway Guide*. London: Harper Collins.

Branca, P., 1975. *Silent Sisterhood: Middle-Class Women in the Victorian Home*. London: Croom Helm.

Brandon, J. A., W. Jones, and M. D. Ohman, 2019. 'Multidecadal Increase in Plastic Particles in Coastal Ocean Sediments'. *Science Advances*, 5(2), eaax0587.

Braun, M. and H. Kingsley (eds.), 2015. *Salt and Silver: Early Photography 1840–1860*. London: MACK.

Bremner, G. A. and J. Conlin, 2015. '1066 and All That: E. A. Freeman and the Importance of Being Memorable'. *Proceedings of the British Academy*, 202, 3–28.

Bridgland, D. R., 1994. *Quaternary of the Thames*. London: Chapman and Hall.

Bridgland, D. R., 2014. 'John Lubbock's Early Contribution to the Understanding of River Terraces and their Importance to Geography, Archaeology and Earth Science'. *Notes and Records of the Royal Society*, 68, 49–63.

Brooks, R., 2009. *Solferino 1859: The Battle for Italy's Freedom*. Oxford: Osprey.

Buckle, H. T., 1858a. *History of Civilization in England*. New York: D. Appleton and Company.

Buckle, H. T., 1858b. 'The Influence of Women on the Progress of Knowledge'. *Fraser's Magazine*, 57, 395–407.

Buckle, H. T., 1859. 'Mill *On Liberty*'. *Fraser's Magazine*, 59, 509–42.

Bulstrode, J., 2016. 'The Industrial Archaeology of Deep Time'. *The British Journal for the History of Science*, 49, 1–25.

Burkitt, M. C., 1923. *Our Fore-Runners*. London: Williams and Norgate.

Burn, W. L., 1964. *The Age of Equipoise: A Study of the Mid-Victorian Generation*. London: Routledge.

Busk, G., 1861. 'On the Crania of the Most Ancient Races of Man, by Professor D. Schaaffhausen, of Bonn. With Remarks, and Original Figures, Taken from a Cast of the Neanderthal Cranium'. *Natural History Review*, 1, 155–76.

Busk, G., 1865. 'On a Very Ancient Human Cranium from Gibraltar'. *Report of the British Association for the Advancement of Science (Bath 1864)*, 34, 91–2.

Bussac, E. de (ed.) 1999. *1859: Naissance de la préhistoire, récits des premiers témoins*. Clermont Ferrand: Éditions Paleo, Préhistoire 3.

Buteux, C.-J., 1857. 'Notions générales sur la géologie du département de la Somme'. *Mémoires de la Société Impériale d'Émulation d'Abbeville*, 8, 561–74.

Bynum, W. F., 1984. 'Charles Lyell's Antiquity of Man and its Critics'. *Journal of the History of Biology*, 17(2), 153–87.

Cannadine, D., 2017. *Victorious Century: The United Kingdom, 1800–1906*. London: Allen Lane, Kindle Edition.

Carroll, L., 1865. *Alice's Adventures in Wonderland*. London: Macmillan.

Caygill, M. and J. Cherry (eds.), 1997. *A. W. Franks: Nineteenth-Century Collecting and the British Museum*. London: British Museum Press.

Childe, V. G., 1958. *The Prehistory of European Society*. Harmondsworth: Penguin.

Chippindale, C., 1988. 'The Invention of Words for the Idea of "Prehistory"'. *Proceedings of the Prehistoric Society*, 54, 303–14.

Christian, D., 2004. *Maps of Time: An Introduction to Big History*. Berkeley, CA: University of California Press.

Christy, H., 1865. 'On the Prehistoric Cave-Dwellers of Southern France'. *Transactions of the Ethnological Society of London*, 3, 362–72.

Clark, J. F. M., 2014. 'John Lubbock, Science, and the Liberal Intellectual'. *Notes and Records of the Royal Society*, 68, 65–87.

Cohen, C., 1998. 'Charles Lyell and the Evidences of the Antiquity of Man', in D. J. Blundell and A. C. Scott, eds., *Lyell: The Past is the Key to the Present*. London: Geological Society Special Publications, 83–93.

Cohen, C. and J.-J. Hublin, 1989. *Boucher de Perthes, les origines romantiques de la préhistorie*. Paris: Belin.

Collini, S., 1991. *Public Moralists: Political Thought and Intellectual Life in Britain, 1850–1930*. Oxford: Clarendon Press.

Collins, W., 1859. *The Woman in White*. London: All the Year Round.

Conan-Doyle, A., 1929. *Sherlock Holmes Long Stories*. London: John Murray.

Conlin, J., 2011. 'An Illiberal Descent: Natural and National History in the Work of Charles Kingsley'. *History*, 96, 167–87.

Conlin, J., 2014. *Evolution and the Victorians: Science, Culture and Politics in Darwin's Britain*. London: Bloomsbury, Kindle Edition.

Cook, J., 1997. 'A Curator's Curator: Franks and the Stone Age Collections', in M. Caygill and J. Cherry, eds., *A. W. Franks: Nineteenth-Century Collecting and the British Museum*. London: British Museum Press, 115–29.

Cook, J., 2012. In pursuit of the unity of the human race: Henry Christy and Mexico, in *Turquoise in Mexico and North America: Science, Conservation, Culture and Collections*, eds. J. C. H. King, M. Carocci, C. Cartwright, C. McEwan & R. Stacey. London: Archetype Publications, 175–80.

Cook, J., 2013. *Ice Age Art: Arrival of the Modern Mind*. London: British Museum.

Corfield, P., 2007. *Time and the Shape of History*. New Haven, CT: Yale University Press.

Croll, J., 1864. 'On the Physical Cause of the Change of Climate during Geological Epochs'. *The London, Edinburgh, and Dublin Philosophical Magazine and Journal of Science*, 28, 121–37.

Croll, J., 1868a. 'On Geological Time, and the Probable Date of the Glacial and the Upper Miocene Period: Part I'. *The London, Edinburgh, and Dublin Philosophical Magazine and Journal of Science*, 35(238), 363–83.

Croll, J., 1868b. 'On Geological Time, and the Probable Date of the Glacial and the Upper Miocene Period: Part II'. *The London, Edinburgh, and Dublin Philosophical Magazine and Journal of Science*, 36(241), 141–54.

Croll, J., 1868c. 'On Geological Time, and the Probable Date of the Glacial and the Upper Miocene Period: Part III'. *The London, Edinburgh, and Dublin Philosophical Magazine and Journal of Science*, 36(244), 362–86.

Daniel, G., 1964. *The Idea of Prehistory*. Harmondsworth: Pelican Books.

Daniel, G., 1972. 'The Origins of Boucher de Perthes' Archéogéologie'. *Antiquity*, 46, 317–20.

Dart, R.A., 1925. Australopithecus africanus: the man-ape of South Africa. *Nature*, 115, 195–99.

Darwin, C., 1859. *On the Origin of Species by Means of Natural Selection, or the Preservation of Favoured Races in the Struggle for Life*. London: John Murray.

Darwin, C., 1871. *The Descent of Man, and Selection in Relation to Sex*. London: John Murray.

Darwin, C. and A. Wallace, 1858. 'On the Tendency of Species to Form Varieties; and on the Perpetuation of Varieties and Species by Natural Means of Selection'. *Zoological Journal of the Linnean Society*, 3, 45–62.

Defrance-Jublot, F., 2011. 'La Question religieuse dans la première archéologie préhistorique 1859–1904', in A. Hurel and N. Coye, eds., *Dans l'épaisseur du temps: Archéologues et géologues inventent la préhistoire*. Paris: Publications Scientifiques du Muséum national d'Histoire naturelle, 279–314.

de Jersey, P., 2008. 'Evans and Ancient British Coins', in A. Macgregor, ed., *Sir John Evans 1823–1908: Antiquity, Commerce and Natural Science in the Age of Darwin*. Oxford: Ashmolean, 152–72.

Desmond, A., 1982. *Archetypes and Ancestors: Palaeontology in Victorian London 1850–1875*. London: Blond & Briggs.

Desmond, A., 1998. *Huxley: From Devil's Disciple to Evolution's High Priest*. Harmondsworth: Penguin.

Desmond, A. and J. Moore, 1991. *Darwin*. London: Penguin.

de Waal, E., 2010. *The Hare with Amber Eyes: A Hidden Inheritance*. New York: Random House.

Dickens, C., 1839 (1978). *Nicholas Nickleby*. Harmondsworth: Penguin.

Dickens, C., 1859 [2000]. *A Tale of two Cities*. London: Penguin.

Dickens, C., 1865 [1912]. *Our Mutual Friend*. London: Thomas Nelson.

Dickens, C. and W. Collins, 1857. 'Lazy Tour of Two Idle Apprentices'., Kindle Edition.

Douglas-Fairhurst, R., 2017. *The Story of Alice*. London: Harvill Secker, Kindle Edition.

Dubois, S., 2011. 'Regards croisés sur deux fondateurs de la préhistoire française: Édouard Lartet et Jacques Boucher de Crèvecœur de Perthes', in A. Hurel and N. Coye, eds., *Dans l'épaisseur du temps: Archéologues et géologues inventent la préhistoire*. Paris: Publications Scientifiques du Muséum national d'Histoire naturelle, 245–66.

Duke of Argyll, 1869. *Primeval Man: An Examination of Some Recent Speculations*. London: Strahan.

Dunant, H., 1862. *A Memory of Solferino*. Geneva: J-G Fick, Kindle Edition.

Duncan, D., 1911. *Life and Letters of Herbert Spencer*. London: Williams and Norgate.

Duncan, I., 1860. *Pre-Adamite Man; Or, The Story of Our Old Planet and its Inhabitants, Told by Scripture & Science*, 3rd edn. London: Saunders, Otley.

Eliot, G., 1859. *Adam Bede*. London: Blackwood, Kindle Edition.

Ellegård, A., 1958 [1990]. *Darwin and the General Reader: The Reception of Darwin's Theory of Evolution in the British Periodical Press, 1859–1872*. Chicago: University of Chicago Press.

Evans, C., 2007. 'The Birth of Modern Archaeology', in D. Gaimster, S. McCarthy, and B. Nurse, eds., *Making History: Antiquaries in Britain 1707–2007*. London: Royal Society of Arts, 185–99.

Evans, C., 2012. 'Archaeology and the Repeatable Experiment: A Comparative Agenda', in A. M. Jones, J. Pollard, M. J. Allen, and J. Gardiner, eds., *Image, Memory and Monumentality: Archaeological Engagements with the Material World: A Celebration of the Academic Achievements of Professor Richard Bradley*. Oxford: Oxbow Books, 297–306.

Evans, J., 1849–50. 'On the Date of British Coins'. *The Numismatic Chronicle and Journal of the Numismatic Society*, 12, 127–37.

Evans, J., 1859–61. 'Thursday, January 19th, 1860 John Evans Esq. F.S.A. Read the Following Communication in Illustration of Numerous Rude Flint Flakes and Implements Exhibited by Mr. Shelley, of Reigate'. *Proceedings of the Society of Antiquaries*, 1, 69–77.

Evans, J., 1860a. 'On the Occurrence of Flint Implements in Undisturbed Beds of Gravel, Sand, and Clay'. *Archaeologia*, 38, 280–307.

Evans, J., 1860b. 'Thursday, June 2nd, 1859 John Evans Read a Paper On the Occurrence of Flint Implements'. *Proceedings of the Society of Antiquaries*, 4, 329–33.

Evans, J., 1863. 'Account of Some Further Discoveries of Flint Implements in the Drift on the Continent and England'. *Archaeologia*, 39, 57–84.

Evans, J., 1865. 'On Portions of a Cranium and of a Jaw, in the Slab Containing the Fossil Remains of the *Archaeopteryx*'. *The Natural History Review*, 12, 415–21.

Evans, J., 1866a. 'On a Possible Cause of Climatal Change'. *Geological Magazine*, 3, 171–4.

Evans, J., 1866b. 'On the Forgery of Antiquities'. *Notices of the Proceedings of the Royal Institution of Great Britain*, 4, 356–65.

Evans, J., 1868. 'On Man and his Earliest Known Works', in *Some Account of the Blackmore Museum*. Devizes, 4–21, 87–103.

Evans, J., 1872. *The Ancient Stone Implements, Weapons, and Ornaments of Great Britain.* London: Longmans.

Evans, J., 1875. 'The Coinage of the Ancient Britons and Natural Selection'. *Notices of the Proceedings of the Royal Institution of Great Britain,* 7, 476–87.

Evans, J., 1882a. 'Unwritten History and how to Read it: I'. *Nature,* 26, 513–16.

Evans, J., 1882b. 'Unwritten History and how to Read it: II'. *Nature,* 26, 531–3.

Evans, J., 1890. *The coins of the ancient Britons: supplement,* London: Quaritch.

Evans, J., 1897. *Ancient Stone Implements, Weapons, and Ornaments of Great Britain,* 2nd edn. London: Longmans.

Evans, J., 1898. 'Presidential Address', in *Report of the 67th Meeting of the British Association for the Advancement of Science, Toronto 1897.* London: BAAS, 3–20.

Evans, J., 1943. *Time and Chance: The Story of Arthur Evans and his Forebears.* London: Society of Antiquaries.

Evans, J., 1949. 'Ninety Years ago'. *Antiquity,* 23, 115–25.

Evans, J., 1955. *The Endless Web: John Dickinson & Co Ltd 1804–1954.* London: Jonathan Cape.

Evans, J., 1956. *A History of the Society of Antiquaries.* Oxford: Oxford University Press.

Evans, J., 1964. *Prelude and Fugue: An Autobiography.* London: Museum Press.

Falconer, H., 1868. *Palaeontological Memoirs and Notes of the Late Hugh Falconer, A. M., M.D. with a Biographical Sketch of the Author.* Volume II. London: Robert Hardwicke.

Figuier, L., 1870. *L'Homme primitif.* Paris: Libraire de L. Hachette et Cie.

Finlayson, C., K. Brown, R. Blasco, J. Rosell, J. J. Negro, G. R. Bortolotti, G. Finlayson, A. Sánchez Marco, F. Giles Pacheco, J. Rodríguez Vidal, J. S. Carrión, D. A. Fa, and J. M. Rodríguez Llanes, 2012. 'Birds of a Feather: Neanderthal Exploitation of Raptors and Corvids'. *PLoS One,* 7, e45927.

Finnegan, D. A., 2012. 'James Croll, Metaphysical Geologist'. *Notes and Records of the Royal Society,* 66, 69–88.

FitzGerald, E., 1859 [1967]. *The Rubayyat of Omar Khayyam.* Kansas City, MO: Hallmark Cards.

Flandreau, M., 2016. *Anthropologists in the Stock Exchange: A Financial History of Victorian Science.* Chicago: University of Chicago Press.

Fleming, J. R., 2006. 'James Croll in Context: The Encounter between Climate Dynamics and Geology in the Second Half of the Nineteenth Century'. *History of Meteorology,* 3, 43–54.

Flower, J. W., 1860. 'On a Flint Implement Recently Discovered at the Base of Some Beds of Drift-Gravel and Brick-Earth at St. Acheul, near Amiens'. *Quarterly Journal of the Geological Society of London,* 16, 190–2.

Fontijn, D. R., 2016. *(Un)familiar and (Un)comfortable—The Deep History of Europe.* Leiden: Universiteit Leiden.

Fowles, J., 1969. *The French Lieutenant's Woman.* London: Jonathan Cape.

Fox-Talbot, W. H., 1844–1846. *The Pencil of Nature.* London: Longmans.

Freeman, E. A., 1867. *The History of the Norman Conquest of England, its Causes and its Results.* Volume 1. Oxford: Clarendon Press.

Frere, J., 1800. Account of flint weapons discovered at Hoxne in Suffolk. *Archaeologia,* 13, 204–5.

Gamble, C. S., 1993. *Timewalkers: The Prehistory of Global Colonization*. Cambridge, MA: Harvard University Press.

Gamble, C.S., 1999. *The Palaeolithic Societies of Europe*, Cambridge: Cambridge University Press.

Gamble, C. S., 2007. *Origins and Revolutions: Human Identity in Earliest Prehistory*. New York: Cambridge University Press.

Gamble, C. S., 2013. *Settling the Earth: The Archaeology of Deep Human History*. Cambridge: Cambridge University Press.

Gamble, C. S., 2014. 'The Anthropology of Deep History'. *Journal of the Royal Anthropological Institute*, 21, 147–64.

Gamble, C. S., J. A. J. Gowlett, and R. Dunbar, 2014. *Thinking Big: The Archaeology of the Social Brain*. London: Thames and Hudson.

Gamble, C. S. and R. Kruszynski, 2009. 'John Evans, Joseph Prestwich and the Stone That Shattered the Time Barrier'. *Antiquity*, 83, 461–75.

Gamble, C. S. and T. Moutsiou, 2011. 'The Time Revolution of 1859 and the Stratification of The Primeval Mind'. *Notes and Records of the Royal Society*, 65, 43–63.

Gardiner, B. G., 2003. 'The Piltdown Forgery: A Re-Statement of the Case against Hinton'. *Zoological Journal of the Linnean Society*, 139(3), 315–35.

Gardner, M., 1965. *The Annotated Alice*. Harmondsworth: Penguin.

Gaudry, A., 1859a. 'Os de cheval et de bœuf appartenant à des espèces perdues, trouvés dans la même couche de diluvium d'où l'on a tiré des haches en pierre'. *Comptes Rendus Hebdomadaires de l'Académie des Sciences*, 49, 453–4.

Gaudry, A., 1859b. 'Sur les résultats de fouilles géologiques entreprises aux environs d'Amiens'. *Comptes Rendus Hebdomadaires de l'Académie des Sciences*, 49, 465–7.

Geikie, J., 1894. *The Great Ice Age and its Relation to the Antiquity of Man*, 3rd edn. London: Stanford.

Gellner, E., 1986. 'Soviets against Wittfogel: Or, the Anthropological Preconditions of Mature Marxism', in J. A. Hall, ed., *States in History*. Oxford: Basil Blackwell, 78–108.

Gill, M., 2009. *Eccentricity and the Cultural Imagination in Nineteenth-Century Paris*. Oxford: Oxford University Press.

Goodrum, M. R., 2009. 'The History of Human Origins Research and its Place in the History of Science: Research Problems and Historiography'. *History of Science*, 47, 337–57.

Goodwin, C. W., 1860. 'On the Mosaic Cosmogony', in *Essays and Reviews*. London: John W. Parker and Son, 207–53.

Gowlett, J. A. J., 2009. 'Boucher de Perthes: Pioneer of Palaeolithic Prehistory', in R. Hosfield, F. Wenban-Smith, and M. Pope, eds., *Great Prehistorians: 150 Years of Palaeolithic Research, 1859–2009*. Lithics, 30, 13–24.

Grayson, D. K., 1983. *The Establishment of Human Antiquity*. New York: Academic Press.

Green, R. L. (ed.), 1953. *The Diaries of Lewis Carroll*. London: Cassell's.

Greene, K., 1999. 'V. Gordon Childe and the Vocabulary of Revolutionary Change'. *Antiquity*, 73, 97–109.

Gruber, J. W., 1965. 'Brixham Cave and the Antiquity of Man', in M. E. Spiro, ed., *Context and Meaning in Cultural Anthropology*. New York: Free Press, 373–402.

Halliday, S., 1999. *The Great Stink of London: Sir Joseph Bazalgette and the Cleansing of the Victorian Metropolis*. Stroud: Sutton Publishing, Kindle Edition.

Hamlin, C., 1982. 'James Geikie, James Croll, and the Eventful Ice Age'. *Annals of Science*, 39, 565–83.

Hammond, M., 1982. 'The Expulsion of the Neanderthals from Human Ancestry: Marcellin Boule and the Social Context of Scientific Research'. *Social Studies of Science*, 12, 1–36.

Harari, Y. N., 2014. *Sapiens: A Brief History of Humankind*. London: Harvill Secker.

Harmand, S., J. E. Lewis, C. S. Feibel, C. J. Lepre, S. Prat, A. Lenoble, X. Boes, R. L. Quinn, M. Brenet, A. Arroyo, N. Taylor, S. Clement, G. Daver, J.-P. Brugal, L. Leakey, R. A. Mortlock, J. D. Wright, S. Lokorodi, C. Kirwa, D.V. Kent, and H. Roche, 2015. '3.3-Million-Year-Old Stone Tools from Lomekwi 3, West Turkana, Kenya'. *Nature*, 521, 310–5.

Harrison, E. R., 1928. *Harrison of Ightham*. London: Oxford University Press.

Hérisson, D., J.-L. Locht, E. Goval, P. Antoine and S. Coutard, forthcoming. 'Archaological Sequences, Framework and Lithic Overview in the Late Middle Pleistocene of Northern France', in B. Scott and A. Shaw, eds., *Peopling La Manche*. Oxford: Oxbow Books.

Hockett, C. F. and R. Ascher, 1964. 'The Human Revolution'. *Current Anthropology*, 5, 135–68.

Hodder, I., 2012. *Entangled: An Archaeology of the Relationships between Humans and Things*. London: Wiley-Blackwell.

Holterhoff, L., 2016. 'Liberal Evolutionism and the Satirical Ape'. *Journal of Victorian Culture*, 21, 205–25.

Hooker, J. D., 1864. 'Note on the Replacement of Species in the Colonies and Elsewhere'. *The Natural History Review*, 4, 123–7.

Horner, L., 1858. 'An Account of Some Recent Researches near Cairo, Undertaken with the View of Throwing Light upon the Geological History of the Alluvial Land of Egypt'. *Philosophical Transactions of the Royal Society of London*, 148, 53–92.

Horrall, A., 2017. *Inventing the Cave Man: From Darwin to the Flintstones*. Manchester: Manchester University Press.

Hublin, J.-J., 1997. 'Boucher (de Crèvecœur) de Perthes, Jacques (1788–1868)', in F. Spencer, ed., *History of Physical Anthropology*. Volume 1 A-L. New York: Garland Publishing, 200–2.

Hughes, K., 2017. *Victorians Undone: Tales of the Flesh in the Age of Decorum*. London: Harper Collins, Kindle Edition.

Hunt, J., 1863a. 'Introductory Address on the Study of Anthropology'. *The Anthropological Review*, 1, 1–20.

Hunt, J., 1863b. 'On the Negro's Place in Nature'. *Memoirs of the Anthropological Society*, 1, 1–64.

Hurel, A., J.-J. Bahain, A. Froment, M.-H. Moncel, and A. Vialet, 2016. 'Retourner à Moulin Quignon'. *L'Anthropologie*, 120, 297–313.

Hurel, A. and N. Coye, 2011. 'Introduction', in A. Hurel and N. Coye, eds., *Dans l'épaisseur du temps: Archéologues et géologues inventent la préhistoire*. Paris: Publications Scientifiques du Muséum national d'Histoire naturelle, 7–37.

Hurel, A. and N. Coye, 2016. 'Moulin Quignon 1863–1864: détours inédits et bilan historiographique'. *L'Anthropologie*, 120, 314–43.

Hutchinson, H. G., 1914. *Life of Sir John Lubbock, Lord Avebury.* Volume. 2. Cambridge: Cambridge University Press.

Huxley, T. H., 1863. *Man's Place in Nature and Other Anthropological Essays.* London: Macmillan.

Huxley, T. H. and L. Huxley, 1900. *Life and Letters of Thomas Henry Huxley.* London: Macmillan.

Imbrie, J. and K. P. Imbrie, 1979. *Ice Ages: Solving the Mystery.* London: Macmillan.

Irons, J. C., 1896. *Autobiographical Sketch of James Croll with Memoir of his Life and Work.* London: Edward Stanford.

Kaniari, A., 2008. 'Evans's Sketches from the Human Antiquity Controversy: Epistemological Proxies in the Making', in A. Macgregor, ed., *Sir John Evans 1823–1908: Antiquity, Commerce and Natural Science in the Age of Darwin.* Oxford: Ashmolean, 257–80.

Kemble, J. M., 1857. 'On the Utility of Antiquarian Collections in Relation to the Pre-Historic Annals of the Different Countries of Europe, with Especial Reference to the Museum of the Academy'. *Proceedings of the Royal Irish Academy*, 6, 462–80.

Kidd, C., 2016. *The World of Mr. Casaubon: Britain's Wars of Mythography, 1700–1870.* Cambridge: Cambridge University Press.

King, C., 2008. 'Evans and the Roman Coinage', in A. Macgregor, ed., *Sir John Evans 1823–1908: Antiquity, Commerce and Natural Science in the Age of Darwin.* Oxford: Ashmolean, 173–88.

Kingsley, C., 1863. *The Water Babies: A Fairy Tale for a Land-Baby.* London: Macmillan, Kindle Edition.

Klein, R.G., 1973. *Ice-age hunters of the Ukraine,* Chicago: University of Chicago Press.

Knappett, C., 2005. *Thinking through Material Culture: An Interdisciplinary Perspective.* Pittsburgh, PA: University of Pennsylvania Press.

Knipe, H. R., 1905. *Nebula to Man.* London: Dent.

Laing, S., 1895. *Human Origins.* London: Chapman and Hall.

Lakoff, G. and M. Johnson, 1980. *Metaphors We Live by.* Chicago: University of Chicago Press.

Lamdin-Whymark, H., 2009. 'Sir John Evans: Experimental Flint Knapping and the Origins of Lithic Research'. *Lithics,* 30, 45–52.

Lane-Fox, A. H. L., 1855. *Instruction of Musketry.* London: Parker, Furnivall and Parker.

Lane-Fox, A. H. L., 1867. 'Primitive Warfare: Illustrated by Specimens from the Museum of the Institution'. *Journal Royal United Services Institution,* 11, 612–45.

Lane-Fox, A. H. L., 1868. 'Primitive Warfare: On the Resemblance of the Weapons of Early Man, their Variation, Continuity, and Development of Form'. *Journal Royal United Services Institution,* 12, 399–439.

Lane-Fox, A. H. L., 1869. 'Primitive Warfare: On the Resemblances of the Weapons of Early Races; their Variations, Continuity and Development of Form; Metal Period'. *Journal Royal United Services Institution,* 13, 509–89.

Lane-Fox, A. H. L., 1876. 'Presidential Address'. *Journal of the Anthropological Institute of Great Britain and Ireland,* 5, 468–88.

Larsen, T., 2014. *The Slain God: Anthropologists and the Christian Faith*. Oxford: Oxford University Press.

Lartet, É., 1865–75. 'On a Piece of Elephant's Tusk, Engraved with the Outline of a Mammoth, from La Madeleine, Dep. Dordogne', in É. Lartet and H. Christy, eds. *Reliquiae Aquitanicae, Being Contributions to the Archaeology and Palaeontology of Périgord and the Adjoining Provinces of Southern France*. London: Williams and Norgate, 168, 206–8, and Plate XXVIII.

Lartet, É. and H. Christy, 1864. 'Sur des figures d'animaux gravées ou sculptées et autres produits d'art et d'industrie rapportables aux temps primordiaux de la période humaine'. *Revue archéologique*, 9, 233–67.

Lartet, É. and H. Christy (eds.), 1865–75. *Reliquiae Aquitanicae, Being Contributions to the Archaeology and Palaeontology of Périgord and the Adjoining Provinces of Southern France*. London: Williams and Norgate.

Latour, B., 2005. *Reassembling the Social: An Introduction to Actor-Network-Theory*. Oxford: Oxford University Press.

Leakey, L. S. B., J. F. Evernden, and G. H. Curtis, 1961. 'Age of Bed I, Olduvai Gorge, Tanganyika'. *Nature*, 191, 478–9.

Lewis, S. L. and M. A. Maslin, 2015. 'Defining the Anthropocene'. *Nature*, 519, 171.

Lightman, B., 2007. *Victorian Popularizers of Science: Designing Nature for New Audiences*. Chicago: University of Chicago Press.

Livingstone, D. N., 2008. *Adam's Ancestors: Race, Religion and the Politics of Human Origins*. Princeton, NJ: Yale University Press.

Locht, J.-L., S. Coutard, P. Antoine, N. Sellier, T. Ducrocq, C. Paris, O. Guerlin, D. Kiefer, F. Defaux, and L. Deschodt, 2013. 'Données inédites sur le Quaternaire et le Paléolithique du nord de la France'. *Revue archéologique de Picardie*, 3(1), 5–70.

Lubbock, J., 1859. 'On the Ova and Pseudova of Insects'. *Philosophical Transactions of the Royal Society of London*, 149, 341–69.

Lubbock, J., 1862. 'On the Evidence of the Antiquity of Man, Afforded by the Physical Structure of the Somme Valley'. *Natural History Review*, 2, 244–69.

Lubbock, J., 1865. *Pre-Historic Times, as Illustrated by Ancient Remains and the Manners and Customs of Modern Savages*. London: Williams and Norgate.

Lubbock, J., 1869. *Pre-Historic Times, as Illustrated by Ancient Remains and the Manners and Customs of Modern Savages*, 2nd edn. London: Williams and Norgate.

Lubbock, J., 1870. *The Origin of Civilisation and the Primitive Condition of Man: Mental and Social Condition of Savages*. London: Longmans.

Lubbock, J., 1872. *Pre-Historic Times, as Illustrated by Ancient Remains and the Manners and Customs of Modern Savages*, 3rd edn. London: Williams and Norgate.

Lubbock, J., 1882. *Fifty Years of Science: Being the Address Delivered at York to the British Association, August 1881*. London: Macmillan.

Lubbock, J., 1887. *The Pleasures of Life, Part 1*. London: Macmillan.

Lubbock, J., 1889. *The Pleasures of Life, Part 2*. London: Macmillan.

Lubbock, J., 1892. *The Beauties of Nature and the Wonders of the World We Live in*. London: Macmillan.

Lubbock, J., 1894. *The Use of Life*. London: Macmillan.

Lubbock, J., 1904. *Free Trade*. London: Macmillan.

Lubbock, J., 1912. *The Origin of Civilisation and the Primitive Condition of Man: Mental and Social Condition of Savages*, 7th edn. London: Longmans.

Lubbock, J., 1913. *Pre-Historic Times, as Illustrated by Ancient Remains and the Manners and Customs of Modern Savages*, 7th edn. London: Williams and Norgate.

Lyell, C., 1830–3. *Principles of Geology, Being an Attempt to Explain the Former Changes of the Earth's Surface by Reference to Causes now in Operation*. Volume 3. London: Murray.

Lyell, C., 1860. 'On the Occurrence of Works of Human Art in Post-Pliocene Deposits'. *Report of the Twenty-Ninth Meeting of the British Association for the Advancement of Science, Notices and Abstracts*, 93–5.

Lyell, C., 1863. *On the Geological Evidences of the Antiquity of Man with Remarks on Theories of the Origin of Species by Variation*. London: Murray.

Lyell, C., 1864. 'Presidential Address'. *Report of the Thirty-Fourth Meeting of the British Association for the Advancement of Science, Notices and Abstracts*, 60–75.

Lyell, C., 1866. *Principles of Geology, Being an Attempt to Explain the Former Changes of the Earth's Surface by Reference to Causes now in Operation*. Volume 2, 10th edn. London: Murray.

Lyell, C., 1872. *Principles of Geology, Being an Attempt to Explain the Former Changes of the Earth's Surface by Reference to Causes now in Operation*. Volume 2, 11th edn. London: Murray.

Lyell, C., 1881. *Life, Letters and Journals of Sir Charles Lyell, Bart*. Volume 2. J. Murray.

Macaulay, T. B., 1848–55. *History of England from the Accession of James II*. London: Longmans.

McBrearty, S. and A. S. Brooks, 2000. 'The Revolution That Wasn't: A New Interpretation of the Origin of Modern Humans'. *Journal of Human Evolution*, 39, 453–563.

MacGregor, A., 2008a. 'Evans and Antiquities from the Roman to the Post-Medieval Period', in A. MacGregor, ed., *Sir John Evans 1823–1908: Antiquity, Commerce and Natural Science in the Age of Darwin*. Oxford: Ashmolean, 131–50.

MacGregor, A. (ed.), 2008b. *Sir John Evans 1823–1908: Antiquity, Commerce and Natural Science in the Age of Darwin*. Oxford: Ashmolean.

MacGregor, A., 2008c. 'Sir John Evans, Model Victorian, Polymath and Collector', in A. MacGregor, ed., *Sir John Evans 1823–1908: Antiquity, Commerce and Natural Science in the Age of Darwin*. Oxford: Ashmolean, 1–38.

MacGregor, N., 2010. *The History of the World in 100 Objects*. London: British Museum Press.

McLennan, J., 1869. 'The Early History of Man'. *North British Review*, 50, 272–90.

MacLeod, R. M., 1970. 'The X-Club: A Social Network of Science in Late-Victorian England'. *Notes and Records of the Royal Society of London*, 24, 305–22.

McNabb, J., 2012. *Dissent with Modification: Human Origins, Palaeolithic Archaeology and Evolutionary Anthropology in Britain 1859–1901*. Oxford: Archaeopress.

McNabb, J., 2015. 'The Beast Within: H. G. Wells, *The Island of Doctor Moreau*, and Human Evolution in the Mid-1890s'. *Geological Journal*, 50, 383–97.

McNabb, J., forthcoming. 'The Beast Without: Becoming Human in the Science Fiction of H. G. Wells', in M. Porr and J. M. Mathews, eds., *Interrogating Human Origins*. London: Routledge, XXXXX.

Malafouris, L., 2013. *How Things Shape the Mind: A Theory of Material Engagement*. Cambridge, MA: MIT Press.

Malane, R. A., 2005. *Sex in Mind: The Gendered Brain in Nineteenth-Century Literature and Mental Sciences*. New York: Peter Lang.

Manouvrier, L., 1908. 'L'Inauguration de la statue de Boucher de Perthes à Abbeville (7 juin 1908)'. *Bulletins et Mémoires de la Société d'anthropologie de Paris*, 9, 539–42.

Mather, J. D. and I. Campbell, 2007. 'Grace Anne Milne (Lady Prestwich): More than an Amanuensis?' *Geological Society, London, Special Publications*, 281, 251–64.

Maxwell, R., 2000. 'Introduction', in *Charles Dickens: A Tale of two Cities*. London: Penguin, ix–xxxiii.

Meek, R. L. (ed.) 1973. *Turgot: On Progress, Sociology and Economics*. Cambridge: Cambridge University Press.

Meek, R. L., 1976. *Social Science and the Ignoble Savage*. Cambridge: Cambridge University Press.

Mellars, P. A., 1973. 'The Character of the Middle-Upper Palaeolithic Transition in South-West France', in C. Renfrew, ed., *The Explanation of Culture Change: Models in Prehistory*. London: Duckworth, 255–76.

Meltzer, D. J., 2009. *First Peoples in a New World: Colonizing Ice Age America*. Berkeley, CA: University of California Press.

Meltzer, D. J., 2015. *The Great Paleolithic War: How Science Forged an Understanding of America's Ice Age Past*. Chicago: University of Chicago Press, Kindle Edition.

Milne, L. E., 1901. 'Memoir', in G. A. Prestwich, *Essays Descriptive and Biographical*. Edinburgh and London: Blackwood, 1–70.

Moncel, M.-H., R. Orliac, P. Auguste, and C. Vercoutère, 2016. 'La Séquence de Moulin Quignon est-elle une séquence archéologique?' *L'Anthropologie*, 120(4), 369–88.

Montagu, A., 1965. *The Human Revolution*. Cleveland, OH, and New York: The World Publishing Company.

Montagu, A., 1972. *Statement on Race*. New York: Oxford University Press.

Moore, J., 1982. 'Charles Darwin Lies in Westminster Abbey'. *Biological Journal of the Linnean Society*, 17, 97–113.

Morgan, L. H., 1877. *Ancient Society*. New York: World Publishing.

Morse, M., 1999. 'Craniology and the Adoption of the Three-Age System in Britain'. *Proceedings of the Prehistoric Society*, 65, 1–16.

Mortillet, G. de, 1867. *Promenades prèhistoriques a l'exposition universelle*. Paris: C. Reinwald.

Mortillet, G. de, 1869. 'Essai d'une classification des cavernes et des stations sous abri fondée sur les produits de l'industrie humaine'. *Comptes Rendus Hebdomadaires des sèances de l'Académie des Sciences*, 58, 553–5.

Mortillet, G. de, 1897. *La Formation de la nation française. Textes, linguistique, palethnologie, anthropologie.* Paris: Alcan.

Moser, S., 1998. *Ancestral Images: The Iconography of Human Origins.* Stroud: Alan Sutton.

Murchison, C., 1868. 'Biographical Sketch of the Author', in *Palaeontological Memoirs and Notes of the Late Hugh Falconer, A.M., M.D. with a Biographical Sketch of the Author.* Volume I. London: Robert Hardwicke, 13–53.

Murchison, R. I., 1808. 'Journal 1861', in *Scientific Correspondence, Notebooks and Diaries Forming the Murchison Collection, in the Geological Society's Archive Collection.* LDGSL/841/25. London: Geological Society.

Murchison, R. I., 1839. *The Silurian System, Founded on Geological Researches in the Counties of Salop, Hereford, Radnor, Montgomery, Carmarthen, Brecon, Pembroke, Monmouth, Gloucester, Worcester, and Stafford; with Descriptions of the Coal-Fields and Overlying Formations.* London: John Murray.

Murray, T., 1992. 'Tasmania and the Constitution of "the Dawn of Humanity"'. *Antiquity*, 66, 730–43.

Murray, T. (ed.), 1999. *Encyclopedia of Archaeology: The Great Archaeologists.* Volume 2. Oxford: ABC-CLIO.

Murray, T., 2009. 'Illustrating "Savagery": Sir John Lubbock and Ernest Griset'. *Antiquity*, 83, 488–99.

Oakley, K. P., 1964. 'The Problem of Man's Antiquity'. *Bulletin of the British Museum (Natural History) Geological Series*, 9, 85–155.

Oakley, K. P., B. G. Campbell, and T. I. Molleson, 1971. *Catalogue of Fossil Hominids. Part II: Europe.* London: British Museum (Natural History).

Owen, J., 2000a. 'The Collecting Activities of Sir John Lubbock (1834–1913): Volume 1', wtheses.dur.ac.uk/1603/: Durham University.

Owen, J., 2000b. 'The Collecting Activities of Sir John Lubbock (1834–1913): Volume 2', wtheses.dur.ac.uk/1603/: Durham University.

Owen, J., 2008. 'A Significant Friendship: Evans, Lubbock and a Darwinian World Order', in A. Macgregor, ed., *Sir John Evans 1823–1908: Antiquity, Commerce and Natural Science in the Age of Darwin.* Oxford: Ashmolean, 206–29.

Owen, J., 2013. *Darwin's Apprentice: An Archaeological Biography.* Barnsley: Pen and Sword Archaeology.

Owen, J., 2014. 'From Down House to Avebury: John Lubbock, Prehistory and Human Evolution through the Eyes of his Collection'. *Notes and Records of the Royal Society*, 68(1), 21–34.

Paddayya, K., 2016. *Robert Bruce Foote and After: A Centenary of Writing Indian Prehistory.* Dehli: Institute of Archaeology, Archaeological Survey of India.

Page, D., 1869. *Chips and Chapters: A Book for Amateur and Young Geologists.* Edinburgh: W. Blackwood and Sons.

Panizzi, A., 1859. *British Museum: A Guide to the Exhibition Rooms of the Departments of Natural History and Antiquities.* London: Printed by order of the Trustees.

Papagianni, D. and M. A. Morse, 2013. *The Neanderthals Rediscovered: How Modern Science Is Rewriting the Story.* London: Thames and Hudson.

Pattison, S. R., 1858. *The Earth and the Word: Or Geology for Bible Students*. London: Longman, Brown, Green, Longmans and Roberts.

Patton, M., 2016. *Science, Politics and Business in the Work of Sir John Lubbock: A Man of Universal Mind*. London: Routledge.

Pauketat, T. R., 2013. 'Bundles of/in/as Time', in J. Robb and T. R. Pauketat, eds., *Big Histories, Human Lives: Tackling Problems of Scale in Archaeology*. Santa Fe, NM: SAR Press, 35–56.

Penck, A. and E. Brückner, 1909. *Die Alpen in Eiszeitalter*. Leipzig: Tauchnitz.

Pengelly, H., 1897. *A Memoir of William Pengelly of Torquay*. London: Murray.

Penwarden, J. and M. Stanyon, 2008. 'The Business Foundation for a Public Career: Evans, the Paper Industry and Life at Nash Mills', in A. Macgregor, ed., *Sir John Evans 1823–1908: Antiquity, Commerce and Natural Science in the Age of Darwin*. Oxford: Ashmolean, 39–50.

Perry, G., 2011. *The Tomb of the Unknown Craftsman*. London: British Museum.

Pettitt, P. B. and M. J. White, 2011. 'Cave Men: Stone Tools, Victorian Science, and the "Primitive Mind" of Deep Time'. *Notes and Records of the Royal Society*, 65(1), 25–42.

Phillips, J., 1860. 'Anniversary Address of the President'. *Quarterly Journal of the Geological Society of London*, 16, xxvii–lv.

Picard, C., 1834–5. 'Notice sur des instruments celtiques en corne de cerf'. *Mémoires de la Société Royale D'Émulation d'Abbeville*, 1834–1835, 94–116.

Picard, C., 1836–7. 'Notice sur quelques instruments celtiques'. *Mémoires de la Société Royale d'Émulation d'Abbeville*, 1836–7, 221–72.

Piggott, S., 1968. *The Druids*. London: Thames and Hudson.

Pitts, M. and M. Roberts, 1997. *Fairweather Eden*. London: Century.

Pizanias, N., 2011. 'Répercussions de l'affirmation de la haute antiquité de l'homme dans la presse chrétienne', in A. Hurel and N. Coye, eds., *Dans l'épaisseur du temps: Archéologues et géologues inventent la préhistoire*. Paris: Publications Scientifiques du Muséum national d'Histoire naturelle, 315–41.

Pole, 1862. 'Photography for Travellers and Tourists'. *Macmillan's Magazine*, 6, 248–51.

Porter, R., 1978. 'Gentlemen and Geology: The Emergence of a Scientific Career, 1660–1920'. *The Historical Journal*, 21(4), 809–36.

Pouchet, G., 1859. 'Hache de pierre trouvée dans le dilivium'. *Comptes Rendus Hebdomadaires de l'Académie des Sciences*, 49, 501–2.

Pouchet, G., 1860. 'Excursion aux carrières de Saint-Acheul'. *Actes du Muséum d'Histoire Naturelle de Rouen*, 1, 33–47.

Prestwich, G. A., 1881. 'Channel Tunnels and Channel Bridges'. *Good Words*, 210–6.

Prestwich, G. A., 1895. 'Recollections of M. Boucher de Perthes'. *Blackwood's Magazine*, 157, 939–48.

Prestwich, G. A., 1899. *Life and Letters of Sir Joseph Prestwich*. Edinburgh and London: William Blackwood.

Prestwich, G. A., 1901. *Essays Descriptive and Biographical*. Edinburgh and London: Blackwood

Prestwich, J., 1850–9. 'Boulogne 1850–59 etc.', in *Prestwich Papers, Field Notebooks*. London: Geological Society Library.

Prestwich, J., 1859a. 'On the Occurrence of Flint Implements, Associated with the Remains of Animals of Extinct Species in Beds of a Late Geological Period, in

France at Amiens and Abbeville, and in England at Hoxne', in *Royal Society Manuscript Collection PT 61 pages 1-116*.

Prestwich, J., 1859b. 'On the Westward Extension of the Old Raised Beach of Brighton; and on the Extent of the Sea-Bed of the Same Period'. *Quarterly Journal of the Geological Society*, 15, 215–21.

Prestwich, J., 1859–60. 'On the Occurrence of Flint Implements, Associated with the Remains of Extinct Mammalia, in Undisturbed Beds of a Late Geological Period [Abstract]'. *Proceedings of the Royal Society of London*, 10, 50–9.

Prestwich, J., 1860. 'On the Occurrence of Flint Implements, Associated with the Remains of Animals of Extinct Species in Beds of a Late Geological Period, in France at Amiens and Abbeville, and in England at Hoxne'. *Philosophical Transactions of the Royal Society of London*, 150, 277–317.

Prestwich, J., 1862–3. 'Theoretical Considerations on the Conditions under which the Drift Deposits Containing the Remains of Extinct Mammalia and Flint Implements Were Accumulated, and on their Geological Age'. *Proceedings of the Royal Society of London*, 12, 38–52.

Prestwich, J., 1863. 'On the Section at Moulin Quignon, Abbeville, and on the Peculiar Character of Some of the Flint Implements Recently Discovered There'. *Quarterly Journal of the Geological Society*, 19, 497–505.

Prestwich, J., 1864a. 'Theoretical Considerations on the Conditions under which the (Drift) Deposits Containing the Remains of Extinct Mammalia and Flint Implements Were Accumulated, and on their Geological Age'. *Philosophical Transactions of the Royal Society*, 154, 247–309.

Prestwich, J., 1864b [1866]. 'On the Quaternary Flint Implements at Abbeville, Amiens, Hoxne, etc; their Geological Position and History'. *Notices of the Proceedings of the Royal Institution of Great Britain*, 4, 213–22.

Prestwich, J., 1873a. 'John Wickham Flower'. *Geological Magazine*, 10, 430–2.

Prestwich, J., 1873b. 'Report on the Exploration of Brixham Cave, Conducted by a Committee of the Geological Society, and under the Superintendence of Wm. Pengelly, Esq., F.R.S, Aided by a Local Committee; with Descriptions of Animal Remains by George Busk, Esq., F.R.S., and of the Flint Implements by John Evans Esq., F.R.S.' *Philosophical Transactions of the Royal Society*, 163, 471–572.

Prestwich, J., 1874. 'On the Geological Conditions Affecting the Construction of a Tunnel between England and France'. *Proceedings of the Institute of Civil Engineering*, 37, 110–45.

Prestwich, J., 1886. *Geology, Chemical, Physical, and Stratigraphical*. Oxford: Clarendon Press.

Prestwich, J., 1888. *Geology, Chemical, Physical, and Stratigraphical*. Oxford: Clarendon Press.

Prestwich, J., 1895. *Collected papers on Some Controverted Questions of Geology*. London: Macmillan.

Ravin, F.-P., 1834–5. 'Mémoire géologique sur le bassin d'Amiens, et en particulier sur les cantons littoraux de la Somme'. *Mémoires de la Société Royale d'Émulation d'Abbeville*, 1834–5, 143–210.

Remy-Watté, M., 2011. '1859 et la naissannce de l'archéologie préhistorique en Normandie', in A. Hurel and N. Coye, eds., *Dans l'épaisseur du temps:*

Archéologues et géologues inventent la préhistoire. Paris: Publications Scientifiques du Muséum national d'Histoire naturelle, 213–44.

Richard, N., 1999. 'Gabriel de Mortillet', in T. Murray, ed., *Encyclopedia of Archaeology: The Great Archaeologists*. Santa Barbara, CA: ABC-Clio, 93–107.

Richard, N., 2002. 'Archaeological Arguments in National Debates in Late 19th-Century France: Gabriel de Mortillet's *La Formation de la nation française* (1897)'. *Antiquity*, 76(291), 177–84.

Richard, N., 2011. 'Les Sociétés savantes et la question de l'antiquité de l'homme', in A. Hurel and N. Coye, eds., *Dans l'épaisseur du temps: Archéologues et géologues inventent la préhistoire*. Paris: Publications Scientifiques du Muséum national d'Histoire naturelle, 267–78.

Rigollot, M.-J., 1854. *Mémoire sur des instruments en silex trouvés à Saint Acheul, près d'Amiens, et considérés sous les rapports géologique et archéologique*. Amiens.

Rigollot, M.-J., 1856. 'Mémoire sur des instruments en silex trouvés à Saint Acheul, près d'Amiens, et considérés sous les rapports géologique et archéologique'. *Mémoires de la Société des Antiquaires de Picardie*, 14, 23–60.

Roberts, A. J. and N. Barton, 2008. 'Reading the Unwritten History: Evans and *Ancient Stone Implements*', in A. Macgregor, ed. *Sir John Evans 1823–1908: Antiquity, Commerce and Natural Science in the Age of Darwin*. Oxford: Ashmolean, 995–113.

Roberts, R., 2016. '"A Wonderful Illustration of Modern Necromancy": W H F Talbot's Promise of Photography', in R. Roberts and G. Hobson, eds., *William Henry Fox Talbot: Dawn of the Photograph*. London: Scala Arts and Heritage Publishers, 11–31.

Roe, D. A., 1968a. 'British Lower and Middle Palaeolithic Handaxe Groups'. *Proceedings of the Prehistoric Society*, 34, 1–82.

Roe, D. A., 1968b. *A Gazetteer of British Lower and Middle Palaeolithic Sites*. London: Council for British Archaeology Research Report 8.

Roe, D. A., 1981. *The Lower and Middle Palaeolithic Periods in Britain*. London: Routledge and Kegan Paul.

Rogers, H. R., 1860. 'The Reputed Traces of Primeval Man'. *Blackwood's Magazine*, 88, 422–39.

Rosny-Aîné, J. H., 1911 [1982]. *La Guerre du feu* (reprinted as *Quest for Fire*). Harmondsworth: Penguin.

Rowley-Conwy, P., 2006. 'The Concept of Prehistory and the Invention of the Terms "Prehistoric" and "Prehistorian": The Scandinavian Origin, 1833–1850'. *European Journal of Archaeology*, 9, 103–30.

Rowley-Conwy, P., 2007. *From Genesis to Prehistory: The Archaeological Three Age System and its Contested Reception in Denmark, Britain and Ireland*. Oxford: Oxford University Press.

Rudwick, M., 2005. *Bursting the Limits of Time: The Reconstruction of Geohistory in the Age of Revolution*. Chicago: University of Chicago Press.

Ruskin, J., 1853. *The Stones of Venice. Volume the Second. The Sea-Stories*. London: Smith, Elder and Company.

Sackett, J., 2014. 'Boucher de Perthes and the Discovery of Human Antiquity'. *Bulletin of the History of Archaeology*, 24(2), 1–11.

Schaafhausen, H., 1858. 'Zur Kenntnis der ältesten Rasseschädel'. *Archiv für Anatomie, Physiologie und Wissenschaftliche Medizin*, 25, 453–78.

Schlanger, N., 2010. 'Series in Progress: Antiquities of Nature, Numismatics and Stone Implements in the Emergence of Prehistoric Archaeology'. *History of Science*, 48, 343–69.

Schlanger, N., 2011. 'Coins to Flint: John Evans and the Numismatic Moment in the History of Archaeology'. *European Journal of Archaeology*, 14, 465–79.

Schneid, F. C., 2012. *The Second War of Italian Unification 1859–61*. Oxford: Osprey.

Secord, J., 2007. 'How Scientific Conversation Became Shop Talk', in A. Fyfe and B. Lightman, eds., *Science in the Marketplace: Nineteenth-Century Sites and Experiences*. Chicago: University of Chicago Press, 23–59.

Sellar, W. C. and R. J. Yeatman, 1931. *1066 and All That: A Memorable History of England Comprising All the Parts You Can Remember Including One Hundred and Three Good Things, Five Bad Kings, and Two Genuine Dates*. London: Methuen.

Serres, M. and B. Latour, 1995. *Conversations on Science, Culture, and Time*. Ann Arbor, MI: University of Michigan Press.

Shackleton, N. J. and N. D. Opdyke, 1973. 'Oxygen Isotope and Palaeomagnetic Stratigraphy of Equatorial Pacific Core V28-238'. *Quaternary Research*, 3, 39–55.

Shea, J. J., 2011. '*Homo sapiens* Is as *Homo sapiens* Was: Behavioural Variability versus 'Behavioural Modernity' in Palaeolithic Archaeology'. *Current Anthropology*, 52, 1–35.

Sherratt, A., 2002. 'Darwin among the Archaeologists: The John Evans Nexus and the Borneo Caves'. *Antiquity*, 76, 151–7.

Shryock, A. and D. L. Smail (eds.), 2011. *Deep history: The Architecture of Past and Present*. Berkeley, CA: University of California Press.

Smail, D. L., 2008. *On Deep History and the Brain*. Berkeley, CA: University of California Press.

Smail, D. L. and A. Shryock, 2013. 'History and the "Pre"'. *American Historical Review*, 118, 709–37.

Smiles, S., 1859. *Self-Help: With Illustrations of Character, Conduct and Perseverance*. London: Harper & Brothers, Kindle Edition.

Smiles, S., 2007. 'The Art of Recording', in D. Gaimster, S. McCarthy, and B. Nurse, eds., *Making History: Antiquaries in Britain 1707-2007*. London: Royal Society of Arts, 123–40.

Smith, W. G., 1888. 'Lepores Palaeolithici: Or, the Humorous Side of Flint Implement Hunting'. *Essex Naturalist*, 2, 7–12.

Snobelen, S. D., 2001. 'Of Stones, Men and Angels: The Competing Myth of Isabelle Duncan's Pre-Adamite Man (1860)'. *Studies in History and Philosophy of Science Part C: Studies in History and Philosophy of Biological and Biomedical Sciences*, 32, 59–104.

Sommer, M., 2004. '"An Amusing Account of a Cave in Wales": William Buckland (1784–1856) and the Red Lady of Paviland'. *British Journal for the History of Science*, 37, 53–74.

Spencer, F., 1990. *Piltdown, a Scientific Forgery*. London: Natural History Museum.

Spencer, H., 1858. *Essays: Scientific, Political and Speculative*. London: Longman.

Steffen, W., J. Grinevald, P. Crutzen, and J. McNeill, 2011. 'The Anthropocene: Conceptual and Historical Perspectives'. *Philosophical Transactions of the Royal Society A: Mathematical, Physical and Engineering Sciences*, 369, 842–67.

Stewartby, L., 2008. 'Evans and the English Coinage', in A. Macgregor, ed., *Sir John Evans 1823–1908: Antiquity, Commerce and Natural Science in the Age of Darwin*. Oxford: Ashmolean, 189–203.

Stocking, G. W., 1971. 'What's in a Name? The Origins of the Royal Anthropological Institute (1837–71)'. *Man*, 6, 369–90.

Stocking, G. W., 1987. *Victorian Anthropology*. New York: The Free Press.

Stringer, C. and C. Gamble, 1993. *In Search of the Neanderthals: Solving the Puzzle of Human Origins*. London: Thames and Hudson.

Sweet, R., 2007. 'Founders and Fellows', in D. Gaimster, S. McCarthy, and B. Nurse, eds., *Making History: Antiquaries in Britain 1707–2007*. London: Royal Society of Arts, 53–67.

Tabrum, A. H., 1910. *Religious Beliefs of Scientists: Including One Hundred Hitherto Unpublished Letters on Science and Religion from Eminent Men of Science*. London: Hunter & Longhurst.

Taylor, R. and E. Wakeling, 2002. *Lewis Carroll Photographer: The Princeton University Library Albums*. Princeton, NJ: Princeton University Press.

Temple, F., 1860. 'The Education of the World', in *Essays and Reviews*. London: John W. Parker and Son, 1–49.

Thackray, J. C., 2004. 'Sir Joseph Prestwich (1812–1896)', in *Dictionary of National Biography Volume 45*. Oxford: Oxford University Press, 276–8.

Thomas, N., 1991. *Entangled Objects: Exchange, Material Culture and Colonialism in the Pacific*. Cambridge, MA: Harvard University Press.

Thompson, M. W., 2009. *Darwin's Pupil: The Place of Sir John Lubbock, Lord Avebury, 1834–1913, in Late Victorian and Edwardian England*. Ely: Melrose Books.

Tilley, C., 1999. *Metaphor and Material Culture*. Oxford: Blackwell.

Toynbee, A., 1884 [2011]. *Lectures on the Industrial Revolution in England: Popular Addresses, Notes and Other Fragments*. Cambridge: Cambridge University Press.

Trautmann, T., 1992. 'The Revolution in Ethnological Time'. *Man*, 27, 379–97.

Trentmann, F., 2016. *Empire of Things: How We Became Consumers, from the Fifteenth Century to the Twenty-First*. London: Allen lane.

Tylor, A., 1863. 'On the Discovery of Supposed Human Remains in the Tool-Bearing Drift of Moulin-Quignon'. *The Anthropological Review*, 1, 166–8.

Tylor, E. B., 1865. *Researches into the Early History of Mankind and the Development of Civilization*. London: John Murray.

Tyndall, J., 1871 [2011]. *Hours of Exercise in the Alps*. Cambridge: Cambridge University Press.

Van Riper, B., 1993. *Men among the Mammoths, Victorian Science and the Discovery of Human Prehistory*. Chicago: Chicago University Press.

Vialet, A., J.-C. Favin Lévêque, M. Lebon, O. Tombret, A. Zazzo, A. Froment, P. Charlier, C. Vercoutère, and A. Hurel, 2016. 'Nouvel Examen des ossements humains de Moulin Quignon (Somme, France). Étude anthropologique, taphonomique et première datation par le radiocarbone'. *L'Anthropologie*, 120(4), 389–427.

Wallace, A. R., 1864. 'The Origin of Human Races and the Antiquity of Man Deduced from the Theory of "Natural Selection"'. *Journal of the Anthropological Society of London*, 2, 158–87.

Wallace, A. R., 1898. *The Wonderful Century: Its Successes and Failures*. Toronto: George N. Morang.

Wells, H. G., 1920. *The Outline of History: Being a Plain History of Life and Mankind*, London: Macmillan.

Wells, H. G., 1921. *The Grisly Folk*. Harmondsworth: Penguin Books.

Wells, H. G., 1922. *A Short History of the World*. London: Cassell's.

Whately, R., 1856. *Miscellaneous Lectures and Reviews*. London: Parker, Son, and Bourn.

White, M. J., 2017. *William Boyd Dawkins and the Victorian Science of Cave Hunting: Three Men in a Cavern*. Barnsley: Pen and Sword.

White, M. J., N. Ashton, and S. Lewis, 2018. 'At the Sign of Handaxe Yard'. *British Archaeology*, 163, 50–5.

White, M. J. and P. B. Pettitt, 2009. 'The Demonstration of Human Antiquity: Three Rediscovered Illustrations from the 1825 and 1846 Excavations in Kent's Cavern (Torquay, England)'. *Antiquity*, 83, 758–68.

Wilson, A. N., 2003. *The Victorians*. London: Arrow Books.

Wilson, B., 2016. *Heyday: Britain and the Birth of the Modern World*. London: Weidenfeld and Nicolson.

Wilson, D., 1851. *The Archaeology and Prehistoric Annals of Scotland*. Edinburgh: Sutherland and Knox.

Wilson, L. G., 1996. 'Brixham Cave and Sir Charles Lyell's ... the Antiquity of Man: The Roots of Hugh Falconer's Attack on Lyell'. *Archives of Natural History*, 23, 79–97.

Wilson, L. G., 2002. 'A Scientific Libel: John Lubbock's Attack upon Sir Charles Lyell'. *Archives of Natural History*, 29, 73–87.

Wilson, P., 1988. *The Domestication of the Human Species*. New Haven, CT: Yale University Press.

Winchester, S., 2011. *The Alice behind Wonderland*. Oxford: Oxford University Press.

Woodward, H., 1893. 'Eminent Living Geologists, No 8 Professor Joseph Prestwich'. *Geological Magazine*, 10(6), 241–6.

Woodward, H., 1899. 'Life and Letters of Sir Joseph Prestwich, M.A., D.C.L., F.R.S., F. G.S., Formerly Professor of Geology in the University of Oxford. Written and Edited by his Wife'. *Geological Magazine*, 6, 373–81.

Worsaae, J. J. A., 1849. *The Primeval Antiquities of Denmark*. London: Parker.

Zilhão, J., 2007. 'The Emergence of Ornaments and Art: An Archaeological Perspective on the Origins of Behavioural "Modernity"'. *Journal of Archaeological Research*, 15, 1–54.

Index

Note: Tables and figures are indicated by an italic "*t*" and "*f*", respectively, following the page number.

For the benefit of digital users, indexed terms that span two pages (e.g., 52–53) may, on occasion, appear on only one of those pages.